ECODESIGN – The Competitive Advantage

ALLIANCE FOR GLOBAL SUSTAINABILITY BOOKSERIES
SCIENCE AND TECHNOLOGY: TOOLS FOR SUSTAINABLE DEVELOPMENT

VOLUME 18

Series Editor   **Dr. Joanne M. Kauffman**
  6–8, rue du Général Camou
  75007 Paris
  France
  kauffman@alum.mit.edu

Series Advisory Board

Professor Dr. Peter Edwards
Swiss Federal Institute of Technology – Zurich, Switzerland

Dr. John H. Gibbons
President, Resource Strategies, The Plains, VA, USA

Professor David H. Marks
Massachusetts Institute of Technology, USA

Professor Mario Molina
University of California, San Diego, USA

Professor Greg Morrison
Chalmers University of Technology, Sweden

Dr. Rajendra Pachauri
Director, The Energy Resources Institute (TERI), India

Professor Akimasa Sumi
University of Tokyo, Japan

Professor Kazuhiko Takeuchi
University of Tokyo, Japan

Aims and Scope of the Series

The aim of this series is to provide timely accounts by authoritative scholars of the results of cutting edge research into emerging barriers to sustainable development, and methodologies and tools to help governments, industry, and civil society overcome them. The work presented in the series will draw mainly on results of the research being carried out in the Alliance for Global Sustainability (AGS).
The level of presentation is for graduate students in natural, social and engineering sciences as well as policy and decision-makers around the world in government, industry and civil society.

For other titles published in this series, go to
www.springer.com/series/5589

Wolfgang Wimmer · Kun Mo Lee
Ferdinand Quella · John Polak

ECODESIGN – The Competitive Advantage

 Springer

Wolfgang Wimmer
Vienna University of Technology
and ECODESIGN company
Vienna, Austria
wimmer@ecodesign.at

Kun Mo Lee
Ajou University
Suwon, Korea
kunlee@ecodesign-company.com

Ferdinand Quella
Siemens AG (retired)
Ottobrun, Germany
qusaeng@t-online.de

John Polak
EnVent
Iroqois, Ontario
Canada
john.polak@envent.ca

ISBN 978-94-007-3348-0   ISBN 978-90-481-9127-7 (eBook)
DOI 10.1007/978-90-481-9127-7
Springer Dordrecht Heidelberg London New York

Printed on acid-free paper

Springer is part of Springer Science+Business Media (www.springer.com)

THE **AGS**
The Alliance for Global Sustainability

Preface

Dealing with environmental issues should no longer be considered simply as a cost of doing business. Effective environmental improvements to a company's products and services can be turned into business opportunities. This book was written with the express purpose of helping managers of companies, in particular of Small to Medium sized Enterprises (SMEs), to better deal with environmental challenges and address customer requirements, all in order to turn their environmental investments into competitive market advantages.

Several examples are provided throughout the book, but also warning signs (Alert Boxes). These "Alerts" are posted to help managers avoid typical traps when working with environmental considerations in business processes.

The authors have many years of experience in the various aspects of implementing Ecodesign. This experience includes working in industry for many years; leading the environmental departments in a multinational company; managing research projects in eco-product development; Life Cycle Assessment; and national and international environmental communication and marketing.

This book is the latest in a series. The 2002 "Ecodesign Pilot" introduced a tool and software to help design more environmentally compatible products. It was directed specifically at designers. The 2004 book, "Ecodesign Implementation", was written to help project managers optimize product development processes from an environmental perspective.

The current book is intended to help CEOs, and management staff, address environmental issues more systematically in their strategic programs and management systems. Although many examples are taken from the electrical and electronics (E&E) industry, managers in all industry sectors will, hopefully, also find the book useful as they also have products, plants, services, and software.

In this book, the authors have attempted to provide managers with some background information on what will be needed, strategic information on what should be done, and advice on different types of tools and actions. A systematic and (hopefully) understandable approach was used in writing the book and includes the kind of information needed at five different decision-making levels:

- Corporate
- Market

- Product
- Production
- Management

These five levels are interlinked with three leading questions:

- What is the problem?
- What should be done?
- How should the problem be approached?

The book starts with a chapter on the potential for environmental issues to contribute to competitive advantage. The concept of Sustainability used is explained and a new business paradigm is defined.

From Chapters 2–4 the five decision-making levels and the three leading questions are combined. The following matrix serves as a roadmap:

	What is it? Chapter 2 – Situation analysis	What should be done? Chapter 3 – Strategy development	How to approach the problem? Chapter 4 – Action plan development
Corporate level	2.1 Corporate analysis	3.1 Corporate considerations	4.1 Corporate action plan
Market level	2.2 Market analysis	3.2 Market considerations	4.2 Market action plan
Product level	2.3 Product analysis	3.3 Product design considerations	4.3 Product design action plan
Production level	2.4 Production analysis	3.4 Production considerations	4.4 Production action plan
Management level	2.5 Management analysis	3.5 Management considerations	4.5 Management action plan

In Chapter 5, three examples are presented:

- CE-Marking of complex set top box
- Redesigning a mobile phone base station
- Carbon footprint, carbon reduction opportunity, and carbon management approach using complex set top box

The book concludes with Chapter 6, in which an outlook on possible future scenarios is explored.

A unique feature of the book is the checklist that accompanies each section. Collectively, these checklists are designed to help managers analyse their situation, develop their strategy, and then develop their action plan for achieving competitive advantage with Ecodesign.

The checklists can be downloaded as files from:

www.ecodesign-company.com/competitive-advantage. The username is "manager", and the password is "ecode$ign".

The checklists in the book and website contain the following elements:

Assessment questions	Answer	Comments	Recommended follow-up activities	Related chapters
Is there a ...?	Yes☺ No☹	Integra, etc.	Opti ...	2.3, 3....

The authors hope that this book will be useful in helping reduce the environmental impacts of a company's products and services, but also in gaining competitive advantage. We encourage the reader to (a) use the ideas described in the book, (b) be aware of the potential traps described in the Alert Boxes, (c) learn from the examples provided, and (d) use the checklists.

Introduction

Environmental Problems

When confronted with a question about their views on the state of our collective environment, many people will likely respond in a way that defines environment in a narrow way, as a uni-dimensional problem, or perhaps as a simple or somehow homogeneous issue. The particular outlook will, of course, vary from person to person, depending on background knowledge, exposure to, and experience with environmental issues. However, it is expected that most would view or define environmental problems as those related to garbage, dirty air, or dirty water. Certainly the routine media reports on smog days, garbage disposal challenges, recycling issues, sewage treatment problems, risks associated with hazardous chemicals, and oil spills would support those views. However, the kinds of environmental problems we deal with today are multi-facetted, non-linear, and complex. We need to understand and deal with them through an open-system approach.

One could argue that there has been an evolution in the kinds of environmental issues we face, from simple and local problems to the current environmental challenges that are much more global and complex in nature. Attempting to deal with all issues in a single way is, of course, naive. On the other hand, attempting to discuss every possible solution for every specific issue is a far too daunting task.

Therefore, it may be easier to think of environmental challenges in three broad categories:

- Direct environmental problems or "end of pipe" pollution types of problems, such as oil spills and raw sewage discharges.
- Indirect environmental problems with unintended consequences or unanticipated effects arising from designed activities. Examples include pesticide development and application leading to impacts on the reproductive capability of waterfowl, sulphur and nitrogen oxide releases from industrial activity leading to acidification of lakes, ozone depletion from releases of chlorofluoro carbons (CFCs), and climate change arising from releases of greenhouse gases.
- Resource mismanagement problems, such as over-fishing, poor forest management practices, and over-consumption of certain metals, that result in scarcity, net resource depletion, and species and habitat loss.

The latter two categories of issues, "Indirect" and "Resource Management" can both be characterized by root causes that were developed as solutions to problems and "seemed like a good idea at the time". DDT (Dichloro-diphenyl-tri-chloro-ethane), CFCs (chloro-fluoro-carbons), PCBs (polychlorinated biphenyls), over-fishing, unsustainable forest management, and gasoline, oil, and coal as primary energy sources, are all examples in which the product, chemical or practice were considered, in their day, to be tremendous progress, if not futuristic breakthroughs. However, no-one anticipated the consequences.

Environmental Implications

When addressing environmental problems at a product level, one should consider all of the environmental issues across the product's life cycle, including the concepts related to the "Carbon Footprint", energy efficiency, opportunities for recycling and reuse, and toxicity. In general terms, this all relates to the "Sustainable Product Development" approach, or, as is intended in this book, "Ecodesign".

And so, Managers, it's time to play a more active environmental role in your company by initiating and promoting the "Ecodesign" of your products. The cost of resources, including energy, and particularly non-renewable resources, is at best unstable, but likely to explode because of either scarcity or over-consumption. "Eco" is no longer some crazy term. In fact, it may well contain the solution to some of your company's challenges and convert these challenges into opportunities.

Exhaustive consumption patterns can be overcome by the application of concepts related to higher levels of sustainability. Recycling of resources not only extends resource life, but also avoids the need for new resources, thereby eliminating the environmental impacts of the related acquisition and refining processes. Environmental strategies are needed, not only for the enterprise, but also for the way the product is designed. It is no longer enough to simply exchange one hazardous material in a product for another that is slightly less toxic. A more sustainable approach uses all of the environmental and cost-saving potential over the entire life cycle of the product.

However, and in particular for Small to Medium-sized Enterprises (SMEs):

- Managers frequently complain that the concepts of sustainability and Ecodesign do not, or perhaps even cannot, be applied to their business. More than 10 years ago, the Boston Consulting Group showed that after Cost and Quality [1], "Environment" has become the most important decision-making criteria for customers. If the environmental properties of a product are superior to those of competitors' products, the rules for that market sector will already have been changed. If the environmental investment has not yet been made, there is still time to become a market leader, even when currently lagging behind.
- It is often thought that an SME need not be concerned about the environment because each individual company sells too few of any particular product each

year to be of broader concern. However, all of the players in a market sector together may well produce millions of similar products. Regulatory agencies may be concerned about the environmental impact of certain types of products in the marketplace, and if adequate environmental progress is not initiated, then new regulations or directives may be imposed. With some degree of new thinking, a leadership role can be attained which, quite possibly, could create a number of opportunities.

− SMEs should not necessarily feel alone in the development of environmental strategies and concepts. They can cooperate successfully in concepts such as "recycling" or "take back" with either other big or small companies or with governmentally organized systems. This works especially well for capital goods where "take back" can be collectively planned, thereby generating direct benefits for customers. Also, recovered parts can often be sold as being functionally as good as new.

− Environmentally compatible products are not necessarily more expensive. By smart management of resources, mass can be reduced and a transition to other materials may be possible. Today, the tight market between suppliers and producers often does not allow companies, especially individual SMEs, to change materials easily. However, co-operation within the supply chain can result in better information about materials and facilitate material substitution.

Another important consideration is that companies do not always just sell hardware products, like computers, in isolation. Products can often play an active role in systems. For example, in the "Internet System", a computer's application and environmental impact will depend on how and where it is used (e.g. private use, while travelling, or in industrial applications). Appropriate environmental planning and preparation can pay significant dividends.

Regardless of physical location, companies must know the local rules and regulations. Even if production takes place in a developing country for a non-regulated global market sector, one should not feel immune from environmental regulations. Many countries are developing regulations related to "take back" requirements, substance restrictions, product labelling and energy consumption. One will need to know what is necessary in the target markets to successfully sell into those markets. Management strategies for environmental risk assessment will help a company to more effectively deal with upcoming regulatory action and even prepare, at least partially, for possible surprises. For example, if a company exports into the European Union, the company should be aware that:

− The European Commission has already announced that companies should not stop with the "Ecodesign Directive" as it is, but extend its application to all kinds of products.

− The "CE" mark ("Conformité Européene" in French) will eventually also guarantee the environmental properties.

− The Integrated Product Policy (IPP) will challenge all manufacturers by providing best practice examples. While the legal future of the recently released Green Paper on IPP is uncertain, the principles contained therein can be expected to be implemented in some manner or another [2].

The "technological age" of a company's product with respect to innovation is another important consideration. Has the company already planned the next "innovation leap"? The competitors, if they haven't done so already, will certainly be thinking about it. Customers will be very interested in new products that, for example, consume 50% less energy.

Addressing the types of problems and implementing the kinds of approaches noted above could very well result in technology and process changes which, in turn, would create better long-term sustainability of resources and, not insignificantly, create immediate and long-term economic benefits. However, it is management's decision whether to be constantly reacting to new and upcoming regulations and market requirements or whether a more proactive approach is desirable in which the environment is viewed as a business opportunity.

Environment as a Business Opportunity

Since the 1960s, awareness of environmental issues has increased dramatically as has the resultant demand for related solutions. In recent years, this has been driven by concern over climate change which has a direct correlation with fossil fuel use and energy efficiency. Pressure in the form of international agreements (e.g. Kyoto Protocol), politics at both the national and international level, non-governmental organizations' calls for action, media coverage, and domestic action plans have all resulted in a high level of public awareness.

Companies, such as General Electric (GE), Philips, Siemens, Walmart, and Tesco have recognized the economic opportunity related to appropriate environmental stewardship and have built it into their corporate strategies through initiatives such as GE's "Ecomagination", and Philips' "Ecovision". A few industries are even calling upon governments to show leadership.

At the same time, and probably not coincidentally, there has been a significant growth in the use of various types of environmental labels, ranging from resource management labels through to labels that consider impacts across the full life cycle. Together with this growth in the use of ecolabels has also come, quite predictably, misrepresentation and misleading or even false environmental claims. While this misuse is certainly regrettable, the overall growth in the use of environmental marketing does signal the importance with which companies (the private sector) deal with the environment overall.

Progressive companies will want to lever their environmental investment to gain market share, in large part through their marketing efforts. On the other hand, some companies are taking a "wait and see" attitude. How will the marketplace respond to these opposing strategies? Will consumers continue to embrace environmental concerns? Will there be a return on environmental investment? Does environmental marketing create further demands that may be difficult to meet?

While one can interpret the demand for environmental action in many ways, the authors contend that it should be seen as a catalyst for economic development. And, furthermore, that this growth already has significant momentum and is not likely to abate for many years to come. The fact is that the multi-dimensional issue of environment has consistently been measured as being in the top three or four concerns for citizens in North America and Europe for the last few decades. Therefore, it should not be surprising that the environment has become a serious component on the political agenda of many countries. In short, this issue of environment is not going away any time soon.

Many signals are already visible. A number of countries are either debating or implementing carbon taxes. Others are considering or actually using carbon trading within regulated cap and trade systems. Many companies and even whole industries are using supply-chain management techniques to not only help them manage financial and quality issues better but also as a mechanism to improve overall environmental management and reduction of environmental impacts across their product's life cycle.

The key point is that, from a business perspective, and the choices that businesses make, the market dynamics of the situation will give rise to opportunities and also threats. The standard "S" Curve of business is likely a good representation of how the marketplace responds to the increase of importance of environment. As the following diagram shows, where innovation takes place, there is rapid growth. As markets become saturated, market maturation is reached and growth levels off. This curve can repeat itself many times in any one sector, with new product development and innovation, or the introduction of a new issue of concern, like environment (see Fig. 1).

Each phase has different characteristics and implications for strategic planning. Of course, the time dimensions and rates of growth are highly variable and depend on the nature of the market and the economic climate in which it operates.

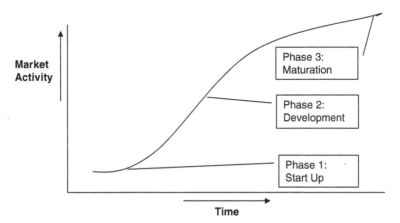

Phase 1 – Start up as the concept has low awareness and acceptance.
Phase 2 – Development as the concept gains higher awareness and acceptance.
Phase 3 – Maturation as the concept reaches full penetration.

Fig. 1 The business S curve

Because of the evolution of the environment as an important issue to consumers, and thus the marketplace, new investment will be attracted, causing new products to be created and bringing in new players. As is normal in any healthy market-based economy, this will create winners and losers. Needless to say, failure to recognize and act in this new context will ensure its own rewards.

Alert

While the discussion above is general in nature, not all product categories are at the same point in the S Curve at the same time. Environmental investments may produce different results depending on the product category. Product specific market analysis from an environmental perspective will pay dividends.

Contents

Chapter 1
Sustainability as a Competitive Business Advantage

"Sustainability", "Sustainable Development", and "Corporate Social Responsibility" are all terms that are increasingly being used in daily communication. Also, companies are finding that the use of environmental claims in their marketing efforts can pay dividends. Product performance related to "lower energy consumption", "better recycling rates", etc., are often declared. But, in considering Sustainability or Sustainable Development from a more holistic perspective, the broader picture needs examination.

1.1 Our Understanding of Sustainability

There are many (perhaps too many) definitions out there about the meaning of Sustainability or Sustainable Development. In order to develop a working understanding, the authors have *not* attempted to come up with more definitions, explaining in comprehensive sentences what cannot be expressed with only a few words.

So we've tried a different approach. Something is definitely not sustainable and cannot be referred to as Sustainable Development if:

- Renewable resources are depleted too quickly (or at all).
- Consumption exceeds need – you would be surprised by the percentage of unpacked food in the waste stream of industrialized countries!
- Products or processes that waste energy (e.g. stand-by consumption).
- Waste products from industrialized countries end up in developing countries for inappropriate "recycling" thereby causing health and environmental damage.
- Companies create environmental hazards or risks or inappropriately exploit staff for enhanced profits, etc.

The list could be quite long and would still be incomplete. Certainly, one can easily understand that the above examples are definitely not examples of sustainability. Although it is difficult to define what Sustainable Development is, it is easier to identify what is just the opposite.

W. Wimmer et al., *ECODESIGN – The Competitive Advantage*,
Alliance for Global Sustainability Bookseries 18,
DOI 10.1007/978-90-481-9127-7_1, © Springer Science+Business Media B.V. 2010

In this book we assume that the reader is aware of the environmental and social problems caused by industrialized countries seeking the "quick and easy dollar or euro" without long-term strategies or thinking. Consequently, we want to focus on the link between Sustainability and the way of doing business.

How Sustainability is relevant to a specific business model requires some exploration. If it is argued that "*this* particular business model is not linked to Sustainability", then the vulnerability of the business model and perhaps the business itself should be assessed. Some questions warrant consideration:

- How will the business be affected if oil costs rise to US$250 per barrel? What will be the direct and indirect consequences?
- What is the most important raw material (e.g. metal) used in the product?
- Where does this material come from?
- How "safe" is the supply chain?
- How scarce are the materials being used in the production of products?

These questions may lead to a new understanding of Sustainability beyond simply being, or being seen to be, "green" or "socially responsible". The concept of Sustainability can serve as a forecasting system to highlight potential problems. The question is how one can use this forecasting system to improve and strengthen a business model to avoid problems in the future – or, even better, to gain competitive advantage because potential problems were detected early and opportunities were acted on accordingly.

Before going further, let's have a look at some developments in the recent past. They are presented here to help predict and understand what the future may bring.

It can be argued that, with the current oil price well under US$100, there is no need to worry about an oil price of US$250. This might be true, but then it may be only a short-term perspective. While there are currently significant oil reserves, the International Energy Agency published the World Energy Outlook (WEO) 2008, which contained some critical warnings. The Outlook noted that: "The world's energy system is at a crossroad. Current global trends in energy supply and consumption are patently unsustainable – environmentally, economically, and socially. But that can – and must – be altered; there's still time to change the road we're on"[1].

In the WEO 2008, the production of 800 oilfields, including the 54 super-giant fields which hold more than five billion barrels of oil, were analysed in detail. The result is frightening: "We estimate that the average production-weighted observed decline rate worldwide is currently 6.7% for fields that have passed their production peak. In our Reference Scenario, this rate increases to 8.6% in 2030." With an annual decline rate of 6.7%, only 50% of the reserves will remain in just 10 short years! With such a dramatic reduction in supply, we can probably guess the impact on price.

Another example is the scarcity of metal resources. The rare metal indium is used primarily in electronic components (especially in lead-free solder), on liquid crystal display (LCD)-glass, and in light emitting diode (LED) screens. Every mobile phone, personal digital assistant (PDA), and Laptop, and flat panel LCD-TV contains a small amount of indium. The world production is around 500 t per year, and occurs predominantly in zinc-sulfide when mining the carrier metal zinc [2].

The worldwide total reserves, including those that are economically extractable today (about 2,800 t), amount to 6,000 t [3]. As such, the supply horizon for this rare metal is only between 5 and 10 years. Business models using new technologies like thin-film solar cells based on indium may be in danger because of the scarcity of this metal.

What does your own business model depend on? Which source of energy? What (scarce) resources? Upon what (currently changing) assumptions or system boundaries, market requirements, etc., are you making plans? With this type of understanding, the concept of sustainability takes a different perspective – it has much to do with addressing business vulnerability. How vulnerable is your business model and are you able to detect early enough where and how to change it according to the main elements of sustainability? The following criteria can be useful guides to follow when designing solutions to problems related to products, services, processes, and technical systems:

- Use of renewable resources
- Use of regional resources
- Level of efficiency in products and processes
- Use of cycles, such as reuse and recycling
- Degree of adaptation and compatibility of technologies
- Level of failure tolerance and risk prevention within systems.

If the analysis and related efforts fail, not only may the business model fail, but there may be broader consequences leading to unsustainable economies.

Nicolas Stern noted in his 2007 report [4] that global warming could shrink the global economy by as much as 20%. But he also stated that taking action against climate change now would cost just 1% of the global gross domestic product (GDP). In more recent statements, he declared that climate change is proceeding faster than expected and that the costs related to dealing with the consequences of global warming now, as opposed to later, are most likely 2% of the GDP.

This raises other questions linked to sustainability:

- How well is your business model prepared to deal with the climate change issue?
- What are the potential consequences of climate change that may affect your business model?

The consequences of climate change could range from direct environmental impacts (storms, floods, etc.) to increased overhead costs (insurance, transportation, energy, emission certificates, etc.), and even to low acceptance of products with large (or not well defined or communicated) carbon footprints.

Throughout this book, the authors have attempted to provide ideas about how to improve the environmental performance of products and whole enterprises, how to lower the vulnerability for your specific business model and, consequently, how to achieve competitive advantage.

We certainly do live in difficult economic times. But we also live in a transitional period, from a fossil energy-based society to a society based on renewable energy and materials. This transition will challenge us in different ways and we will need to find new and smart approaches with which to move forward. The old approaches and pathways will no longer be useful. From a business perspective, as well as for the environment on which we rely, it may well be too dangerous and too expensive to hold on to the old ways much longer.

Meeting the future challenges using an Ecodesign approach aimed at sustainable product and service development also helps move towards more sustainable business models. We may not know exactly how and where this may lead, but what is known is that business models ignoring sustainability will be in real danger. Consequently, managers developing business models and engineers developing products and technical systems need to integrate sustainability issues now. But how does product sustainability translate into your business model? What does this mean for your own products? The following sections lay out what the authors see as important environmental and sustainability trends to be aware of as you design the future of your own products and your business.

1.2 European and Worldwide Regulations and Standards

There are a variety of environmental trends that have the potential to dramatically influence the acceptance, and thus the design, of products. These trends relate to legislative and regulatory initiatives, standards, evolving programs, and market demands. Every manager should have in place a process to identify, analyze, and evaluate existing and forthcoming legislation, regulations, standards, and other product-related programs and pressures. This is important not only in the country of primary market activity but also in the major economies around the world.

At the moment, the product-related environmental issues that are subject to probable regulatory action include

- Management systems
- Ecodesign
- Reduction of GHG emissions, as well as improving energy efficiency
- Substance restrictions
- Take-back and recycling
- Batteries
- Packaging.

Existing environmental product legislation can often be obtained from manufacturers' associations [5] that get information on their products of interest from legal consultants or through their network of political organisations around the world. Specialized consultants also frequently offer investigations into, or regular reports about, forthcoming legislation. Individual producers should then analyze these reports in order to determine which of the restrictions are most applicable.

Often, a restriction starts in one country and then spreads. Typically before any such restriction is established as legislation or regulation, proposals are usually published, public consultation is invited and then after some time, perhaps even years, the restrictions are enacted into law. For example, the first drafts of the RoHS (Restriction on Hazardous Substances) directive [6], were published about 10 years before the legislation was put into force. Nonetheless, many companies seemed somehow surprised by the eventual enactment and, as a result, were negatively affected financially because of their late development of material substitutes.

As the draft documents become available, they should be evaluated and progress followed. In Europe, emphasis is placed on the "precautionary principle" to avoid future problems, while in the Americas, the emphasis has been more frequently on prosecution after damage occurs.

When a new product-related environmental law is put into force, action on existing and newly developed products will be required. In the meantime, the restrictions, especially those related to lead-free soldering, have indeed spread through many parts of the world, causing the technology to change, even in those countries where lead for soldering was not restricted.

Alert

The lack of legal restrictions (environmental or health related) on a particular substance in one country does not guarantee that there will not be legal action against the company if the substance is restricted or otherwise regulated in other countries where the company is also active. A prominent example is asbestos which is still used in many countries of the world.

In Europe, there is considerable emphasis on product-related environmental legislation. This book will not deal with this in much detail. However, in the following subsections, a synopsis related to the new Ecodesign directive concerning Energy Related Products (ERP) [7] and *Climate Change* are provided as examples. Definition of ERP see Annex 8. The new Ecodesign directive extends the well-known EuP directive where only Energy using Products were affected. Now additionally to EuP Energy Related Products are included in the scope.

1.2.1 Ecodesign Directive and Implementing Measures

The Ecodesign directive is a framework directive based on a "new approach" [8]. Only a basic framework of the requirements is included in the directive and the details are described in the "Implementing Measures" which are developed by the

European Commission for specific product groups such as motors and drives, televisions, set top boxes, refrigerators, street-lights, etc.

There are two major goals embedded in the Ecodesign directive related to the attainment of sustainable development (i) resource conservation and (ii) improvement in energy efficiency of energy-related products. Energy efficiency and Ecodesign have been chosen as the two major pillars to achieve these goals. Energy efficiency is aimed at mitigating GHG emissions to reduce climate change effects, while the Ecodesign is aimed at realizing integrated product policy or Sustainable Consumption and Production (SCP) [9].

At the moment, the Ecodesign directive is applied to products that have a significant sales volume (i.e. more than 200,000 units per year) in the EU market. The directive will impose generic, as well as specific, Ecodesign requirements. Generic requirements include those based on the ecological profile of an ERP without setting limit values for selected environmental parameters. Specific requirements are in the form of limit values for selected environmental parameters. In the implementing measure, specific requirements are listed as Ecodesign requirements and generic requirements usually fall under conformity assessment.

Energy efficiency is currently the focal point of the Ecodesign directive. Since most companies have devoted their efforts to enhancing the energy efficiency of their products during use, this issue is no stranger to manufacturers. However, manufacturers are experiencing some difficulties in meeting the generic Ecodesign requirements because of a general lack of experience in the concepts and methodologies of Ecodesign. These frequently relate to the selection of parameters for product redesign and in choosing design options that lead to the improvement of the product's environmental performance.

The purpose of assessing the product is to identify significant environmental aspects throughout its lifecycle. When environmental assessment is carried out, manufacturers can set up the ecological profile which lists environmentally relevant product characteristics and parameters that can be expressed in measureable physical quantities. The level of desired performance is specified under the specific Ecodesign requirements. A systematic Ecodesign process provides the basis for meeting the generic requirements.

Conformity assessment determines, directly or indirectly, whether relevant requirements are fulfilled. Typical activities include: sampling, testing and inspection, evaluation, verification, registration, and accreditation.

The manufacturer's obligation for compliance with the Ecodesign directive includes assessing environmental aspects of the product (and developing an ecological profile), designing and developing products by meeting generic and specific Ecodesign requirements, and carrying out conformity assessment for the CE marking of the product. Section 5.1 shows a relevant CE marking example [10].

For conformity assessment of an ERP, the manufacturers can choose between the internal design control, set out in Annex IV, and the management system set out in Annex V of the Ecodesign directive. As proof of conformity to the Ecodesign directive, the CE mark must be affixed before the product is placed in the EU market. In imposing threshold limit values, no distinction is made between types of the same product, such as the different types of televisions now available (LCD, PDP, rear projection, CRT).

Alert

Are you aware that if your product does not meet the requirements of the Ecodesign directive your product can be removed from the marketplace? Without a CE marking, a product cannot be sold in the European Union.

Integrated Product Policy (IPP) is the next Ecodesign legislation that is being tested by the EU Commission through pilot projects [11].

Commission consultants, together with industry experts, are examining what environmental improvements are possible over the life cycle of predetermined product types. These improvements are then expected to become benchmarks for all manufacturers of similar products. One example of such an investigation involves environmental improvements of cell phones (Commission consultants, together with Nokia) [12].

Given the range of programs aimed at climate protection around the world, a number of restrictions can be expected. The following subsections describe what is currently being planned to reduce GHG emissions.

1.2.2 Climate Change

Climate change is a major challenge faced by mankind. Pressure from the implementation of the Kyoto protocol and Copenhagen Accord will force industry to take action to mitigate GHG emissions from their operations as well as their products and services. Climate change is mainly linked to energy consumption, and electronic products account for a significant portion of this consumption. As such, these products will be targeted by the GHG reduction regulations. Disclosure will be required on the level of GHG emissions and on mitigation efforts. For some, the climate change issue is a threat, while for others it can become an opportunity.

The Kyoto protocol was adopted by the United Nations Framework Convention on Climate Change [13] at its third conference of the parties in 1997 in Kyoto, Japan. It is based on the principle of "common but differentiated responsibilities". The core of the Kyoto protocol is that developed countries should reduce GHG emissions by an average of 5.2% from 1990 levels during the period between 2008 and 2012. Flexible mechanisms were established by the Kyoto Protocol to facilitate meeting the GHG emissions reduction Target: They are the Clean Development Mechanism (CDM) [14] the Joint Implementation (JI) [15], and Emissions Trading [16].

A new global climate treaty, the Copenhagen protocol was not adopted in Copenhagen, Denmark 2009, but is now addressed for Cancun, Mexiko for 2010.

In the so-called Copenhagen Accord it was only decided to restrict global warming to 2C [17].

Key supporting tools will include pricing pollution by introducing a global emissions trading system. This system will be supported by other policy instruments, such as energy efficiency standards, building standards, consumer information requirements, new forms of regulation, and securing significant investment for the energy infrastructure in developing countries through an improved CDM.

The cap-and-trade system [18] will have major impact on many industries, including the electronics sector. The system is based on the allocation of a cap on emissions over a period of time. The authority allocates allowances, and each allowance represents a defined emissions amount (e.g. ton of CO_2 equivalent). It creates scarcity in the market, generating a positive value for the allocated emission amount. Thus, companies must reduce GHG emissions from their operation, their organization, and from their products. The target in the EU is to reduce GHG emissions by 20% from 1990 levels by 2020. Roadmaps are already available [19] which allow anyone to estimate the influence or impact of their own product group. One early result is the restriction on incandescent light bulbs in the EU, which began, on a phased basis, in 2009.

In order for companies to better tackle GHG emission objectives, an accurate estimation of GHG emissions from the entire life cycle of a product is necessary. The GHG emissions data can be used for the identification of those areas requiring improvement. The improvement areas would most likely be those parts, materials, and/or processes that have a high relative contribution to GHG emissions of the product. When a product is redesigned, the design should address the opportunities to reduce GHG emissions throughout the entire life cycle of the product.

The EU uses several tools to address a certain measure or product related restriction:

1. Setting limits by law
2. Voluntary agreements by industry
3. Standards mandated by the EU Commission, but developed by industry and interested parties

Voluntary agreements, in this sense, are not actually as voluntary as is often believed because the agreements are signed and so act as commitments or covenants. An example is the voluntary agreement to use only one kind of power supply for all cell phones. This agreement was implemented successfully only after the EU Commission applied considerable pressure on the parties to sign the agreement.

1.2.3 Standardization such as IEC 62430 and Others

Standardization in Europe and China has, in many cases, attained the characteristic of legislation. For ERP, all interested parties can participate in the related European working groups which have mirror committees at the International Electrotechnical Commission (IEC) level (IEC technical committee (TC) 111) and at the national level.

Most important is the standardization of Ecodesign procedures through IEC 62430 which is explained below.

1.2.3.1 IEC 62430 – Environmentally Conscious Design (ECD)

Differing Ecodesign methods can be a challenge to the fulfillment of legal requirements such as those related to the Ecodesign directive. Conformity assessment methods mandate that manufacturers disclose Ecodesign procedures in their technical documentation in prescribed ways (i.e. according to Annexes IV or V of the Ecodesign directive).

In order that the Ecodesign method should not be considered a technical barrier to trade, the method must be harmonized in the EU as a minimum, but preferably internationally. In this regard, the EU mandated the Comité Européen de Normalisation Electrotechnique (CENELEC) (European Committee for Electrotechnical Standardization) to develop an Ecodesign standard for the implementation of the Ecodesign directive, CENELEC brought the standardization task to the international arena through IEC. In 2005, IEC created TC 111, the environmental technical committee, which launched the environmental standard development projects. The ECD standard IEC 62430 was published in 2009.

The key point of IEC 62430 is the Ecodesign process for the conformity assessment of products, either for the internal design control or the management system. In either case, the same Ecodesign process applies. The Ecodesign process consists of

(i) Analysis of the regulatory and stakeholders' environmental requirements
(ii) Identification and evaluation of environmental aspects and corresponding impacts
(iii) Design and development
(iv) Review and continual improvement

The Ecodesign process steps can fit into the Plan-Do-Check-Act (PDCA) cycle of a management system. Steps (i) and (ii) correspond to *Plan*, step (iii) to *Do*, and step (iv) to *Check* and *Act*. From steps (i) and (ii), significant environmental parameters are identified. These parameters become the target for the environmental improvement. A design solution to improve the environmental performance of the parameter is developed during the design and development step. Once the environmentally improved product is developed and put on the market, any feedback from the market is reflected to the redesign of the product for further improvement of the environmental performance.

IEC 62430 allows for conformity assessment of the Ecodesign requirements, which was not possible with the former ISO TR 14062 "Management systems – design and development of environmentally compatible products". The latter will be redesigned into an ISO Guide.

In IEC TC 111 the "testing methods for the RoHS substances" have been published as international standards in 2009. In addition, the "material declaration" is under development as international standard. Standardization of "Stand-by", "energy management", and "product carbon footprint" will be considered as future projects and will have worldwide implications. The US "Energy Star" standards are also influential as they continuously publish limits on energy consumption of electronic products and have become the basis for legislation in many countries.

Also, the forthcoming International Organization for Standardization (ISO) 26000 "Corporate Social Responsibility (CSR)" standard will have many additional environmental requirements. Only a few of them were originally mandated at the European level but many of them will likely also become part of international legislation, certifications, or purchasing agreements. For these reasons, environmental standards should be treated with the same level of importance as legislation. An interesting observation, and potentially a problem, is that environmental standards are developed from different Technical Committees of IEC and ISO.

Within ISO, the ISO 14001 standard [20] has become the globally accepted standard for environmental management systems, and was based on the Eco-management and Audit Scheme – commonly known simply as EMAS [21].

For environmental communications and labels, ISO has developed the 14020 series of standards: 14020 for all types of environmental labels [22], 14021 is for self-declarations [23], ISO 14024 [24] is for life cycle based environmental leadership (e.g. European Flower, Canadian Environmental Choice, Nordic Swan, German Blue Angel), and 14025 [25] is for life cycle based information declarations (Environmental Product Declarations).

The Japanese "Top runner" approach is a combination of benchmarking with the best-in-class product and continuous improvement [26]. After every improvement in the best-in-class product, the target for the other products is raised.

The procedure for Life Cycle Assessment (LCA) is delineated in the ISO 14040 series of standards [27].

IEC and ISO standards on energy management are also on the horizon. Those standards, which do not become part of legislation, often become requirements in the national and international competitive purchasing processes. Companies not able to fulfill these types of "standards" requirements frequently cannot compete and therefore lose potential business.

Standards can also serve as an opportunity for industry to define what may be missing in new legislation and may actively be driven worldwide to avoid discrepancies in trade conditions. For many new laws, standards for "How to measure?" or "How to do?" are required and should be expected. The next likely candidate for the development of new standards may well be "Registration, Evaluation, Authorisation and Restriction of Chemicals (REACH)" where test procedures are currently missing.

1.3   Forecast of Environmental Trends

In general, the following trends can be expected:

For *management systems*

- More management systems are probable, like ones for Ecodesign and energy management. Integration in one general system is another trend.

For *Ecodesign*

- Energy efficiency of products will be a major issue in the future. There will be more products regulated by specific requirements, and not only those covered by the current Ecodesign directive.
- Benchmarks will be set by the IPP projects to initiate voluntary changes, particularly by the market leaders.
- More environmental and social labelling for products can be expected.

For *Greenhouse gas control*

- Climate protection programs, and reduction targets combined with restrictions for certain products will be implemented.
- Standardization of communicating the carbon footprint of products is underway [28].

For *Energy consumption*

- The Energy end-use efficiency and energy service directive will require reduction of energy consumption by 1% per year for 6 years or more [29].
- The new EU action plan, which runs until 2020, has a goal of 20% reduction in energy consumption, and 20% of energy sources coming from renewable energy.
- Smart grids can be expected.

For *Substance restrictions*

- Further RoHS bans can be anticipated [30] on lead, mercury, cadmium, brominated biphenyls and diphenyl ethers, chromate(VI). Modification of the initial RoHS regulation will include medical equipment and automation and control instruments in its regulatory scope. Consideration will be given to substances such as Bisphenol A, and all brominated flame retardants in the near future.
- Investigation of a further 30,000 "old" substances will be undertaken by REACH [31].

For *Take back and recycling*

- WEEE directive [32] will be more widely implemented. It will be extended by an updated Waste Framework directive [33] requiring reuse of components.
- Take back systems will be implemented on a much wider basis – companies will understand the value of the resources trapped in discarded products.
- Batteries: Initiatives will be taken in relation to waste collection systems, reduction of hazardous substances in batteries.
- Packaging: Reusable packaging will be required for the purpose of waste reduction.

For *other regions around the world*, it is expected that further EU regulations will be adopted in many major countries, although not always in a completely harmonized fashion.

Standards and legislation can also play an important role for Green Procurement in helping customers (buyers of all types) and suppliers (manufacturers and distributors) have a common understanding upon which to build requirements.

1.4 Finding the New Business Paradigm

When consumers (individual, institutional, and corporate) begin buying "green" in sufficient numbers, the marketplace for the affected products becomes transformed.

The request for "green" or even "sustainable" products is already "out there". It is no longer acceptable for large organizations to ignore the environmental performance of or the social implications caused by products they are buying. For example, assume that a large bank or insurance company committed to Corporate Social Responsibility wants to buy new office furniture. They usually buy large quantities and these days they will ask critical questions:

- What is the environmental/CO_2 performance of the equipment we are going to buy?
- Are there social standards guaranteed for the products we buy?
- Are there any toxic/problematic substances present in the product?
- What is the guaranteed recycling rate for the product at the end of its life – or is take-back provided by the manufacturer?

These and other questions will most probably need to be answered to even qualify for consideration for any contract. It is all about honest communication and an honest way of dealing with sustainability issues.

- *Green Procurement*

Through green procurement, the environmental product leaders will gain a market share at the expense of those with lesser environmental profiles. Those products losing a market share will either continue the erosion in their market position or make sufficient environmental investments (best done through proper Ecodesign techniques) to either regain their lost market share, or, as a minimum, stop the erosion or loss of further market share. In this sense, green procurement is a very potent market mechanism for improving our collective environment.

However, there are some complications for the supplier. What kind of environmental attributes and profiles do the consumers of greener products want? What approach should the supplier take in providing product related environmental information? These questions will be dealt with later in Chapter 4. For now, it is important to understand the degree to which environment is becoming important in your customers' purchase decisions.

The green purchasing movement is bounded on one side by the average consumer who is increasingly concerned about all of the environmental impacts of his/her activities. On the other side is the corporate and industrial buyer who regards the environment from two perspectives:

(i) Effective supply chain management, which includes requirements on the environmental performance of the upstream suppliers and their products.
(ii) The downstream customers who may well have environmental requirements or values of their own. These requirements could be driven by end-consumer concerns, environmental lobby forces, or by ever more stringent regulatory pressures.

These two factions (the consumer and the corporate/industrial buyer) tend not to be well organized amongst themselves. However, the same is not true of the middle faction which includes a whole range of institutional, commercial, and retail buyers and who are increasingly aware of the environmental benefits they can generate through their purchasing activities. In many regions of the world, they have realized the value of coordinating their environmental efforts. As a result, more formal organizations have been created to help them. Green Purchasing Networks have popped up in Europe, North America, and Asia, and an International Green Purchasing Network has been launched to provide a more global source of information and coordination.

These networks (or information multipliers) evolved as an efficient way of sharing experience and information without having to "reinvent" the process each time. Some examples:

- The North American Green Purchasing Initiative (NAGPI) is, in effect, a network of networks involving buyers, sellers, certifiers, and information multipliers:

  (a) Buyers and sellers together form the marketplace.
  (b) Certification organizations provide credible standards and information on environmental leadership.
  (c) Green Purchasing Networks serve to promote green procurement and provide linkages and information to their selected audiences.

  In this model, the NAGPI serves as a coordinating body for all the participants and interested parties, and as clearing house of information, through its information hub or web site [34].

- The International Centre for Local Environmental Initiatives (ICLEI), in its dual function as a sustainability clearinghouse and a network of local authorities, has assumed the mandate to coordinate European sustainable procurement initiatives at the local level. ICLEI's BIGNet [35] provides local authorities throughout Europe with experience and information on sustainable procurement and fosters cooperation amongst public authorities.

- Japan's GPN [36] was established to promote green purchasing in Japan. At the moment it had over 3,000 members made up of private sector companies, local authorities, consumer and environmental groups, and cooperatives. The GPN prepares purchasing policies, guidelines for various products, and maintains a product database. It also holds training sessions and awards commendations to organizations that have demonstrated exemplary green purchasing practices.

These Green Purchasing Networks can be very useful to manufacturers, distributors, and suppliers in helping them understand the environmental needs of their various audiences and clients.

Alert

Be wary of assuming what might or might not be acceptable as "green" by consumers or institutional buyers. Unless checked, one may not know what issues they deem to be important or what pressures they might be under.

• *Sustainable Consumption and Production (SCP)*

Partly driven by Green Procurement, the whole life cycle of products and services will need to be made more sustainable. The target of Ecodesign must be Sustainable Consumption and Production (SCP), and involve all of the participants in the different life cycle stages. SCP is a concept that was formally discussed in 1992 at the Rio Conference and was specifically identified as an area for action in "Agenda 21" as a way of addressing product related environmental impacts across the product life cycle.

With all of the regulatory and market pressures, the challenge being faced by suppliers and manufacturers today is how to integrate sustainability into industrial policy and operations. Two major contributing factors are (i) the cost of energy, and in particular, the cost of fossil fuels, and (ii) the awareness of the need for a low carbon future. Both of these factors are moving manufacturers towards more sustainable consumption and production.

While sustainability and environmental management were considered a cost of doing business less than a generation ago, they have become important parts of business and marketing strategy as well as corporate valuation. In many ways, effective sustainability and environmental management have become market opportunities and elements in determining corporate value.

Many countries, companies, and even industry associations, not just in Europe, but around the globe are developing Action Plans for SCP. Typically, the Action Plans include consideration of

• Related regulatory requirements and policy trends in target markets and manufacturing sites.
• Innovations in environmental management and strategy.
• Life cycle assessment.
• Opportunities to implement the six elements of eco-efficiency [37].

  ○ Reduction of material intensity.
  ○ Improvement in energy usage, in particular, efficiency gains across the product life cycle.
  ○ Reducing risks to the environment and human health, primarily through the reduction in the use and release of toxic materials.

- ○ Finding ways to reuse products or components thereof and recycle materials.
- ○ Conserve resources by focusing on the nature and renewability of the energy and materials needed for a product or process.
- ○ Design products such that the product delivers more service for a given amount of environmental inputs. In particular, this means increased durability, repairability, upgradability, multifunctionality, as well as shared use.

- Environmental and energy labeling of products.
- Examination of opportunities for industrial ecology or industrial symbiosis, wherein waste products from one operation or company form the feedstock for another company or process.
- Application of Environmental Management Systems such as ISO 14001 or Europe's EMAS.
- Application of Life Cycle Costing (LCC) [38] approaches to find out where costs occur for customers, the environment, and for business. Many so-called "design-to-cost" approaches take a short-sighted view and frequently focus too much on the capital cost alone. LCC is a methodology for obtaining more sustainability by considering the costs of consumables, repair, reuse, maintenance and service, and disposal.

In summary, considerable activity is taking place worldwide on the subject of Sustainable Consumption and Production eventually leading to a new business paradigm.

- *The Service Business*

A new business paradigm, rising from the sustainability issue, calls for a complete overhaul of the existing business paradigm. The major difference being that a new paradigm not only aims to maximize profits in the short term, but includes long-term (strategic) thinking that understands that

- If there is a severely damaged environment, it will become difficult to run any kind of business in a successful way.
- If climate change continues at the current pace, the market will punish those that contribute the most.
- If more and more customers realize that the products they buy cause severe problems to other humans, they simply won't buy those products.
- With a smart approach, significantly reducing environmental burden, improving social standards, and having a successful business will not be mutually exclusive.

How can all that be possible?

Well, let's say you are a producer of paint and you sell the paint to someone who manufactures steel doors. Let's further assume there is a constructive supplier-manufacturer relationship. Both parties are doing well and business is good.

You are the supplier – environmental questions about how you produce your paint have never been an issue. Your manufacturing facility meets all of the relevant environmental regulations.

But then, one day the door manufacturer comes and tells you that his clients want to know about the environmental performance of the doors he is delivering to them and, as such, needs to know details about your paint products and related manufacture. Certainly, this may be worrisome – should you really provide all the information, some of which may be delicate? But can you really risk losing this business partner?

In a conventional business paradigm, the environmental information would simply be added to the list of requirements and that's it – everyone is happy.

But – the new sustainability driven business paradigm might develop a different approach, with the main issue being the decoupling of your own profit from the amount of products sold.

Let's see if this might be possible.

Up until now, your old business model has worked well. If you sold more and more paint, you earned more and more money. So far, so good. However, for every additional kilogram of paint you sold to your customer, environmental problems have increased.

In the new business paradigm, decoupling the amount of products sold from profits, one can make it work in such a way that even if you sold less paint, you could earn more.

This is only possible if you can get involved in the painting process at the door manufacturer. You take over – simply because you have the expertise of not only producing but hopefully also of applying the paint. Suddenly you need to define a new "product" for which you can be paid. Previously, it was the kilogram of paint – now it is the square meter of correctly painted surface.

The door manufacturer – the expert in handling steel, and not necessarily the expert in handling paint, is now benefiting through a guaranteed quality of the painted surface for a predefined price. You as a paint producer benefit from all the creative ideas related to reducing the amount of paint and solvent to achieve the previously agreed surface quality of your customer's doors. And last, but not the least, the environment is benefiting from fewer resources used, less energy being consumed, and less solvents being used, etc.

You, as a previous paint producer, have become a "service provider" – you provide now the service of delivering painted surfaces at a pre-defined level of quality.

This kind of "*service*" approach is gaining momentum as the basis for viable business models. Servicing refers to the selling of the service or function offered by a product instead of selling the product itself.

Think about your heating (or cooling) or even lighting system in your home. How would that change if the electricity supplier did not sell electricity, but rather sold conditioned indoor air or brightly lit rooms?

You may think that this is not possible. However, these types of models have already been developed. In existing business models, the user, or customer, owns the product. However, in the new business model, the manufacturer or service provider retains ownership of the product while selling the related services to the customer through either lease or rental arrangements.

Leasing of a photocopier is a successful example of the new business model, where the business's service provider – in this case the photocopier producer – makes a profit from the arrangement and the users are free from maintenance headaches while getting a quality copy function. In this paradigm, it is in the best interests of both parties to have a photocopier that requires little maintenance, has reduced costs of production and disposal, while ensuring a high level of quality service (i.e. copying). In this approach, there are incentives to minimize the cost of the photocopier and to extend its useful life through to the eventual reuse of its component parts. This includes consideration of the use of common parts, easy to disassemble connections, the use of environmentally benign materials, and easy product disassembly. The operating cost of the photocopier should also be a consideration at the design stage as consumables, including energy, during use is a major issue for customers.

This example illustrates how a successful business model contributes to the reduction of resource consumption and emissions and, if well done, earns the loyalty of customers. Other well-known business models include closed-loop recycling of onetime use products, such as single use cameras. More innovative business models are expected to emerge as more companies look for opportunities to remain competitive in the era of sustainability.

Sustainable consumption and production can be achievable if and only if the societal paradigm on consumption changes toward the realization of sustainable development. This will involve changes in the consumption habits of consumers and in the business paradigm of the product manufacturers and service providers. The newly developed business paradigm aims at achieving sustainable development. As such, it can be coined the "sustainable business paradigm" or model.

At the base of the needed societal and business paradigm changes, lies the product. The more sustainable product will be less resource consumptive, and generate fewer emissions than the average competing product. This means that the environmental performance – including resource use – of the product needs to be continually improved in order to meet the sustainable business model requirements. The type of product that fits the model may be termed a "sustainable product". In essence, the sustainable product is an "ecoproduct".

- *Product Life*

One of the major factors governing the revenue and cost of the old business model is the useful life of the product. Product life is defined as the length of time the owner of the product owns and is able to make use of the product. Revenue is generated to the producer when the new product is sold. This occurs when either a new product type is introduced or when an already owned product is no longer in use or has been discarded. Design strategies such as reuse can only be possible when the old product is available. Thus, understanding the lifetime of a product is necessary for the Ecodesign of a product.

There are two different types of product lifetimes: Physical or technical, and value or useful lifetime.

- The physical lifetime is the time between product purchase, or when the product is put into service, and final disposal or when it breaks down or deteriorates such that it can no longer deliver its intended function.
- The value lifetime has the same start point (i.e. purchase) and end point (i.e. when the product is discarded), but the reason for disposal is unrelated to the usefulness or usability of the product. In this type of situation, the product no longer meets the requirements or expectations of the owner, even though it still retains some amount of physical lifetime.

Disposal of an electronic gadget or some fashion product is, in most cases, related to the value lifetime, while disposal of durable machinery probably happens at the end of the machine's physical lifetime.

For disposal after a physical lifetime, function depletion and failure of the product to deliver its intended function are the main causes. For disposal after a value lifetime, appearance, capacity and size, and value deterioration are the main causes.

Function depletion in products such as ball point pens, paper, diapers, etc., are examples of products where the main functions have been depleted or totally consumed. Failure occurs when a product fails to deliver its intended function due to physical reasons.

Appearance can be the cause of disposal in many types of products. Capacity and size can be the cause in the case of children's clothes or computer monitors. Value deterioration can occur because of obsolescence of the product due to rapid change in technological terms or technological innovation such as that experienced with IT equipment (e.g. mobile phones and personal computers).

Reuse and recycling are indispensable factors to consider in product design for the new business model, especially closed-loop recycling. In this business model, a prerequisite to reuse and recycling is the availability of disposed products for the manufacturing of new products. The new products should be designed in order to be able to use the parts and/or components recovered from the disposed products. This means that both manufacture and disposal of the same type, or even the same model of the product, must occur in a given time period. If this is not the case, for example for durable home appliances such as washing machines, the business model must be modified to allow overlapping of manufacture and disposal. If overlapping occurs, then the degree of overlap should be increased to maximize the reuse potential. The reuse and recycling should also include paying due consideration to the collection and remanufacturing or refurbishing processes of the product.

The New Business Model

Each individual company must develop its own business model to be competitive in its market sector or segment. The company needs to

- Identify the relevant market trends
- Determine how to adapt its product to these trends
- Decide on a realistic business model

- Determine the performance specifications of the product for the business model
- Implement Ecodesign to develop the product

In developing the business model, the reason for disposal of the product should be identified by analyzing the causes. In implementing Ecodesign, the product designer must consider overlapping the manufacturing and disposal durations before incorporating reuse strategies into the product design.

New environmentally based opportunities can arise for all kinds of business. For example:

(a) A producer of *capital goods*, like medical equipment, can easily include "take back" and "reuse" in his business activity. Refurbishment can become a new business line (profit centre). For capital goods in particular, energy consumption can often be significantly improved by applying energy saving components [39].
(b) A company active in the *maintenance* or general *service* sectors can reuse components as spare parts. Leasing models can be added to get products back in time for reuse. Remote control enables longer service intervals.
(c) If you are developing and producing *software* look at the energy consumption caused by your software programs that it requires less loading commands of batteries or stand-by (cf. checklist in Annex 2).
(d) More and more hardware is also now replaced by software solutions.

While individual companies can change their business paradigms, co-operation with standardization organizations and industry associations can often be helpful in overcoming the broader resistance to change. In the following example, a standard was created to overcome the problem of low customer trust in refurbished and reused goods. This was helpful in changing the market and for helping meet some related waste regulations.

Example

With IEC 62309 a standard was created which defines a *new product* as one which could contain used components "qualified as good as new". This standard is directed to a producer of mass products who can include such "qualified as good as new" components in his production chain. Many used components do not differ from new ones, some cannot be distinguished. A potential between 15% and 25% of the volume in the average could be reused. The standard describes a quality and communication concept.

Also, very old technologies, like vacuum cleaning, can be improved by investigating the behaviour of the people doing the cleaning (see the following example).

Example

When investigating vacuum cleaning, one should examine how such vacuuming is actually done. Improvements could be made to the usability of the cleaner, the service (exchange of bags) and energy utilization.

Many more opportunities for improvement can be identified when the application, or actual use, of products is investigated. In the next example, a PC, in its application of the Internet, can have many consequences. Only a few are mentioned, but the huge environmental reduction potential is demonstrated.

Example from the Internet

One PC has a low impact, but all of the PCs on the Internet represent millions of energy consumers. The number of cellular phones sold annually will become billions. New applications, like telephoning directly via the Internet, increases energy consumption considerably. The growth of the Internet is exponential and this growth produces secondary consequences such as delivery of a package ordered via Internet and a range of social behavioural changes which may have a strong impact on the environment.

One last point. A market leader can set environmental and other market trends. Followers can become best in the market with new innovative products. Besides quality and costs, environmental properties can often influence purchase decisions. Where to position the company is an important question for management. By adopting a more environmental or sustainable business paradigm, even a small company can become a market leader. Small companies can also cooperate with bigger ones, for take back systems or for new product developments if their own resources are not sufficient.

Alert

It is absolutely necessary for SMEs to co-operate with competitors and suppliers in order to change production processes such as changing from lead soldering to lead-free soldering. Otherwise there would significant cost disadvantages for an SME to make this kind of change alone. The same is valid for the development of more compatible technologies. Also, information about regulatory trends is better gathered in cooperation. SMEs can get direct information about forthcoming regulation and influence it by becoming a member of a manufacturers organization.

Also, companies should be cautious about over-promoting small environmental gains. This kind of advertising can back-fire.

1.5 Getting Assistance

Companies considering the possibilities and opportunities arising from Ecodesign might feel alone with such a complex subject. Happily, many countries have started project networks and have found ways to provide funding for environmentally challenging, yet interesting projects. In the EU, such funding is available for special development and information projects related to the Ecodesign directive. Other funding opportunities may be available for climate protection programs. With national funding, similar projects are also promoted, and partnerships often make it easier for funders to provide approvals.

As the procedures vary from country to country, we cannot offer specific recommendations. However, program and contact details in all cases are most likely available from either the national industry associations or the national government agencies responsible for environment or industrial affairs. One long-standing program in Austria can be seen at `http://www.nachhaltigwirtschaften. at/english/index.html`.

Financial assistance may be available for a range of projects and support, including training, formation of special expert networks to exchange experience, or for co-operation in development projects for processes, technologies or completely new products. Often, companies are also assisted in creating concepts for the development of new, more environmentally compatible products.

Chapter 2
Situation Analysis

In the previous chapter, some definitions and examples of the concept of sustainability and probable trends in environmental matters and regulations were presented. Chapter 2 will serve as a framework to help analyze a company's current situation and identify strengths and weaknesses. The result of this situation analysis will be the basis for the strategy development in Chapter 3.

The situation analysis will follow the logic model of (i) corporate, (ii) market, (iii) product, (iv) production, and (v) management. The checklists at the end of each section are designed to help a company undertake its own analysis.

At the end of this chapter, one should have a much better understanding of the potential business effects of sustainability.

2.1 Corporate Analysis

The first section of this chapter on Situation Analysis examines the overall sustainability of a company at the corporate level. At this level, decisions influencing the entire company are made, in particular those that relate to corporate direction and major investment. Thus, it is important to define the boundaries within which environmental improvements are possible.

If no strategic direction is provided nor commitment made to move towards sustainability through, for example, ecoproduct development, there can be only limited scope and freedom for related decisions by managers responsible for marketing, product development and/or manufacturing. But, if environmental initiative is taken and support provided by corporate management, ecoproduct development can become an important driver for environmental progress and competitive market advantage.

As a first step, a series of questions are asked in order to determine the current environmental status of the company. This should help prepare corporate-level management for related strategy development and decision making in Section 3.1.

W. Wimmer et al., *ECODESIGN – The Competitive Advantage*,
Alliance for Global Sustainability Bookseries 18,
DOI 10.1007/978-90-481-9127-7_2, © Springer Science+Business Media B.V. 2010

Organization and corporate culture

First and foremost, the code of conduct of the company should be examined:

- Has a code of conduct already been developed?
- If so, does it include the concept of sustainability?
- To what degree does the company's actual behaviour reflect the code of conduct?
- What yet needs to be done to ensure that behaviour reflects the code of conduct statement?

The existence of an environmental management system and the way of organizing and evaluating processes (design and manufacturing processes, responsibilities) should be analyzed next. Is a review and audit plan of environmental management available? Is an assessment of potential environmental risks along the value chain available?

Production

For production, it is important to understand whether core processes are clearly defined and how the manufacturing processes are structured. What efforts have been made to increase energy and resource-use efficiency in the various manufacturing stages? Has ISO 14001 or an Eco-Management and Audit Scheme (EMAS) been introduced in the company?

Does the company manufacture the main components of its products itself or are they purchased from external suppliers? What is the in-house core competence?

Product

Is Ecodesign part of the management system?

For the company's products, it is important to know their environmental performance. Are Life Cycle Assessment (LCA) results for products available? Are product and production related environmental data available?

How much effort has been put into significantly improving the energy and resource efficiency of the products? How well is the product performing when compared with the best available technologies?

Benchmarking the company's products against competing products helps to understand the key environmental performance indicators such as:

- Level of energy consumption
- Level of material consumption
- Degree of modularity
- Relative content of renewable materials
- Number of reusable parts and components
- Recycling rates
- Toxicity of components and perhaps of the whole product

Benchmarking serves to identify and quantify the degree of innovation potential available for the product.

Supply chain and logistics

Integrating suppliers, as well as managing and improving the relationship with them, are important initiatives in ensuring the supply chain remains robust. As such, it is important to know if auditing programs for suppliers are in place and whether these also include the social aspects of sustainability? Are green purchasing guidelines available?

Where are the suppliers located and how is transportation organized?

Communication

The external perception of any company depends, to a large extent, on its attitude towards outreach and communication. Several relevant reporting mechanisms have been developed and are available to support the kind of corporate image a responsible company may seek. Has the company made a commitment toward public reporting on corporate sustainability and has a Corporate Social Responsibility (CSR) report been made available?

Are other forms of environmental reporting and environmental communication available – especially those about products? Is the process of development of Environmental Product Declaration (EPD) common practice for all of the company's new products?

Staff

The concept of sustainability developed at the corporate level should be implemented and "lived" by all staff members. Therefore, responsibility for all parts of the product life cycle/value chain needs to be assigned – who will be the lead agents for the different environmental issues?

Additionally, regular awareness training on environmental and sustainability issues are valuable.

With the following checklist, the current status of the company at the corporate level can be determined, potential weak points identified for follow-up, and references made to corresponding chapters.

Checklist for corporate analysis

Assessment questions	Answer	Comments	Recommended follow-up activities	Related sections
Do you have an environmental management system in place?	Yes ☺ No ☹	ISO 14001, EMAS	International extension	2.4, 3.5
Is Ecodesign already included in a management system?	Yes ☺ No ☹	IEC 62430, Ecodesign directive Annex IV/V, ISO 9001 or ISO 14001, integrated management system	Review	2.3, 3.5

(continued)

Checklist for corporate analysis (continued)

Assessment questions	Answer	Comments	Recommended follow-up activities	Related sections
Are the missing environmental processes identified and installation planned?	Yes ☺ No ☹	Energy, recycling, communication, chemicals, software	Continuous extension	2.3, 2.4, 3.5
Are the missing environmental tools identified and application planned?	Yes ☺ No ☹	LCA, EPD, design rules, product carbon footprint (PCF), environmental quality function deployment (EQFD)	Extension	2.3, 3.3
Are environmental responsibilities checked at all levels and are responsible persons assigned?	Yes ☺ No ☹	Responsibility at the management level to determine level of expert assistance required	Responsibilities for new processes	2.3, 2.4, 3.5
Is a document available describing all environmental tasks, responsibilities, and processes?	Yes ☺ No ☹	Environmental principles are also a guide for customers and public	Annual check of validity	2.3, 2.4, 3.5
Is quality management involved in checking environmental attributes of components, materials, and products?	Yes ☺ No ☹	Incoming tests required, specifications required, integration in software systems, contracts required	Update regularly	1.2, 2.3, 2.4, 3.5
Have sources been identified for systematic information about trends in legislation and new standards?	Yes ☺ No ☹	Sources could be associations, consultants, government web sites	Continuous participation	1.2
Has risk management been installed for dealing with environmental risks?	Yes ☺ No ☹	Includes trend analysis of legislation, standards, competition	Annual update	1.2, 2.3, 2.4, 3.5
Is an early-warning system in place to (a) identify environmental problems with components or materials in time and (b) to inform management?	Yes ☺ No ☹	Problems can come from suppliers and public media. Authorized persons must be available to manage crises	Test system	1.2, 3.5

2.2 Market Analysis

Corporate environmental stewardship is one of a few key trends shaping the business world of the future. Led by mainstream Fortune 500 companies, corporate environmental management is good business. It enhances share value and market penetration. It will require tools, some of which are already available on the market, such as Ecodesign. This wave of environmental protection is a key trend for tomorrow's business.

Evidence of this trend includes:

- The development of the Dow Jones Sustainability Group Index (DJSGI) which tracks roughly 225 companies that are considered leaders in their sectors in sustainable practices and operations. Notwithstanding the recent volatility of the markets, the DJSGI has outpaced the performance of comparable equity indices such as the Dow Jones Global Index.
- The firms identified by DJSGI are far from being considered peripheral players in the global economy. Rather, the firms identified as leading the global shift toward corporate environmental stewardship and sustainability include mainstream performers such as BMW, Bristol-Myers Squibb Co., Dofasco Inc., Fujitsu Ltd., Procter & Gamble Co., Dow Chemical Co., Siemens, Philips, and Unilever N.V. All are mainstream organizations that have targeted sustainability as core success values for the future.
- In a Harvard Business Review article, W. Chan Kim (Boston Consulting Group) and Renee Mauborgne (INSEAD Distinguished Fellow) described "environment friendliness" as one of the "six levers" for effective and foresighted business planning in the twenty-first century (the others were more traditional: customer productivity, simplicity, convenience, risk mitigation, and fun and image) [1].
- When the Global Environmental Management Initiative (an NGO of Fortune 500 companies) interviewed 30 leading multinationals (mainstream companies like Canon, Compaq, Motorola, Weyerhaeuser, GM, and Du Pont), it found that all had environmental management systems of some shape, and that all included environmental procurement as part of that system.
- Heinz recently acquired Earth's Best Baby Foods in order to access the environmental market demographic. For the same purpose, Colgate-Palmolive acquired Murphy's Oil Soaps. Similarly, recognizing the market opportunity for environmental services, Price-Waterhouse Coopers recently bought EcoBalance, a global environmental consulting practice.

None of this corporate initiative is about environmental altruism alone. It's about how today's most foresighted companies are sowing the seeds for commercial success in the future. There is no current dispute: corporate sustainability and prolonged shareholder value is based, in part, on environmental stewardship. This is one of the waves shaping the business world of the next 25 years.

In any assessment of the marketplace, a number of considerations warrant examination. The most fundamental are (i) the nature of the product itself and (ii) the sector of the market in which that product operates (including the views and perspectives of the customers of that sector of products).

More specifically:

(i) From a product perspective, there are some useful questions to think about when considering how to proceed:

- From your customers' perspective, is the product purchased on its own merits or the services the product provides? For example, an automobile provides transportation for people, but is most often purchased for a wide range of unrelated reasons – status, power, colour – to name a few. This question becomes important when developing a marketing strategy.
- What are your customers' environmental sensitivities or attitudes to your product category? For example, people often think of recycled content when discussing paper products and many will think about the potential health impacts of cleaning chemicals. Similarly, most products which consume energy in the use stage of their life cycle, will create concerns about the costs of energy consumption and perhaps even those impacts related to climate change. These customer concerns may be driven by internal factors, or by outside factors related to export markets or other market pressures. In any case, customers expect that suppliers actually think in advance, thus helping them (the customers) avoid potential problems.

(ii) From a market sector perspective, there are similarly some useful questions to review as part of any analysis:

- What is the size and level of competition in the market sector in which you operate?
- To what degree is your market sector segmented? A sector can have multiple segments as well as different types of customers, all of which behave in different ways. For example, in the cleaning products sector, there are many different types of cleaning products available, and also a variety of end-consumers. Those could relate to individual households, institutional cleaning (janitorial) companies, or commercial/industrial operations.
- What are the environmental trends acting upon or influencing your market sector and what is the prognosis for those trends in terms of future directions and strength? The trends may relate to a variety of factors:

    (a) Governmental directives and regulations such as the European Union directives on Ecodesign and RoHS
    (b) Consumer concerns over toxic chemicals or food safety, or
    (c) Environmental campaigns about specific issues, such as forest management or climate change

- How many companies are competing in your market sector, and how many of them are already active in the area of environmental marketing?
- If your competitors have been active in using environment in their marketing efforts, what types of environmental messaging approaches are they using, and to what degree have their efforts been well received by their customers and the general public?
- Given the current environmental issues acting on your product sector, to what degree should you be prepared to consider all of the environmental impacts of your product over its life cycle? This should relate to the strength of the current environmental trends and the likelihood of customer questions or demands that go beyond the environmental issues currently being addressed.

- What are the most effective channels, vehicles, and information systems available to best position your product from a performance and environmental perspective? Section 4.2 will help identify and explain the options available and which might be best suited.

Answering these questions will be helpful in determining what type of environmental communication to use and how best to position your product.

Checklist for market analysis

Assessment questions	Answer	Comments	Recommended follow-up activities	Related sections
Do your customers show interest in environmental/ sustainability issues yet?	Yes ☺ No ☹	If yes, there is opportunity. If no, you will need to be prepared to communicate your product's environmental attributes soon	Conduct market analysis Prepare framework of your environmental/ sustainability communication	3.2, 4.2
Do your competitors use environment in their marketing?	Yes ☺ No ☹	If no, there is opportunity to lead. If yes, you will need to consider your options	Develop plan for own environmental communication	1.1, 1.3, 2.3
Are you able to effectively communicate the environmental benefits of your products yet?	Yes ☺ No ☹	Evaluate the environmental impacts of the key life cycle stages and key environmental performance indicators of your products	Determine the environmental profile of your products and examine communication options	1.1, 3.2, 4.2
Do you currently use environmental communications in your marketing?	Yes ☺ No ☹	If yes, make sure your claims are correct, trustworthy and reliable. If no, use it as an opportunity	Evaluate your current environmental communication and marketing and extend it with a product perspective	4.2

2.3   Product Analysis

The development of any ecoproduct will require focus on four major aspects: (i) environmental, (ii) resource use efficiency, (iii) degree of environmental hazard, and (iv) the degree of extended material utilization.

In assessing the environmental aspects, consideration of product and stakeholder requirements are necessary. Analysis of the efficiency of resource use will require an analysis of the possibilities for, and ease of, recycling. The aspects related to environmental hazards will require an examination of all of the critical substances, with special attention given to the toxicity of materials and components. The extended material utilization aspects will require analysis of the potential to use recycled and more eco-friendly materials.

2.3.1 Environmental Aspects

Environmental aspects can be defined as elements of the processes, parts, materials, or activities of a product that have the potential to interact with the environment. In this sense, interaction refers to the use of input resources, the generation of emissions and any products, by-products, or co-products that become outputs. Environmental aspects are, by definition, broad in scope. As such they are not quantifiable. However, environmental parameters, relating to the environmental aspects, are quantifiable. Examples of environmental parameters include:

– The type and quantity of materials used (weight, volume)
– Power consumption
– Emissions
– Rate of recyclability
– The time required for disassembly

A more exhaustive list of such parameters can be found in the references [2].

The IEC defined environmental parameters as "quantifiable attributes of an environmental aspect" [3]. Thus, any one environmental aspect can be understood as a collection or category of environmental parameters. For instance, emissions to the atmosphere would be considered to be an environmental aspect, while specific emissions (e.g. CO_2, NO_x) would be considered to be environmental parameters.

Ecodesign aims at improving the environmental performance of a product by reducing the environmental impact of significant environmental parameters. Thus, the presence or absence of significant environmental parameters can be used as indicators for the evaluation of the environmental performance of a product.

The analysis of the environmental aspects of a product should be approached from two different perspectives: (i) the product life cycle and (ii) stakeholder requirements. The former can be envisaged as those environmental aspects that are internal to the manufacturer of a product, while the latter as those aspects that are external to the manufacturer. The analysis of the product life cycle is the identification of significant environmental aspects and thus significant environmental parameters of a product throughout its entire life cycle. The analysis of stakeholder requirements includes the identification of significant environmental parameters from the various external pressures on the product and its manufacturer, such as legal requirements, ecolabelling criteria, and competitors' product performances.

Once a product has been chosen for the environmental improvement process, the product will need to be clearly described in terms of environmental parameters. The purpose of describing a product this way is twofold: (1) to define the reference point, or starting point, in order that improvements resulting from the Ecodesign can be identified; and (2) to identify the significant environmental parameters which are targeted for improvement.

Next, a system boundary of the product system should be defined. This is termed product modeling. The product system boundary encompasses all of the life cycle stages of the product. In each life cycle stage, relevant values of the pre-selected environmental parameters are determined.

The environmental aspects of the product life cycle can then be analyzed by using either a Life Cycle Thinking (LCT) matrix tool or Life Cycle Assessment (LCA) tool, depending on the needs of the user. The environmental aspects of the stakeholder requirements can be analyzed using such tools as Environmental Quality Function Deployment (EQFD) or Environmental Benchmarking (EBM). The significant environmental parameters can be identified from both analyses [2].

2.3.1.1 Analysis of the Product Life Cycle

Life cycle thinking implies consideration of all relevant environmental aspects of a product in its entire life cycle. The aim of the analysis is to identify significant environmental parameters of a product. Here, due attention should be paid to consider all relevant life cycle stages.

There are various analytical tools available. Most well known are LCA and the matrix tools such as the LCT matrix, among others. The former is a comprehensive tool. However, it takes much effort and time to implement. The latter is simple to use but it lacks rigour in analysis.

The LCT matrix tool is described in the Ecodesign standard of electrical and electronic products, IEC 62430 [3]. The method consists of columns and rows, where lifecycle stages and environmental aspects are listed, respectively. For a cell where the relevant life cycle stage and environmental aspects meet, relevant environmental parameters with their values are recorded. Table 2.1 shows an example of an LCT matrix used in the LCT matrix tool [28].

Once all of the relevant environmental parameters with values are collected and the cells of the matrix filled in, their values need to be converted into values with common units such as CO_2 equivalents. The purpose of the conversion is to calculate the relative contribution of each environmental parameter to the product system. This calculation can only be made when all values of the environmental parameters in the LCT matrix are expressed in the same units. The conversion simply involves the multiplication of the environmental parameter values in each cell by the corresponding CO_2 equivalents. Dividing the CO_2 equivalent value of each environmental parameter by the total CO_2 equivalent value of the product, gives the relative contribution of

Table 2.1 Life cycle thinking matrix

Life cycle →	Use of raw				End of
Environmental aspect↓	materials	Manufacturing	Distribution	Use	life
Raw material and energy consumption					
Emissions to air, water and soil					
Physical pollution					
Waste material					
Reuse, recycling, and recovery of material and energy					
Total					

each environmental parameter. By comparing the magnitude of each parameter's relative contribution, the significant environmental parameters can be identified. Similarly, the most significant environmental aspects and life cycle stages can also be identified.

The analysis of the product life cycle by LCA has already been well documented [2]. As such, the use of an LCA tool for the identification of significant environmental parameters is not discussed here. Although the LCA tool is different in depth and breadth from the LCT matrix tool, the outcome remains largely the same. In the case of LCA, however, a more complete ecological profile (cf. Section 5.3) of a product can be generated so that, for example, the carbon footprint of the product can be more readably obtained.

Alert

LCA data can only be compared globally if they are measured with the same physical units. Local or regional data should only be compared on a local or regional level. Only scientifically world-wide accepted impact categories should be used, such as "global warming" or "ozone depletion". Many impact categories used today are not in accordance with ISO 14040. In addition, the weighting of impacts of a product often results in summaries of non-comparable effects. Such results should be evaluated with extreme care.

2.3.1.2   Analysis of Stakeholder Requirements

Stakeholder requirements include the needs of:

(i)   The end users
(ii)   The business to business (B2B) customers

(iii) Environmental directives and regulations
(iv) Eco-labels (voluntary product criteria to achieve certification)
(v) Standards (general/procedural requirements to fulfil)
(vi) Competitors (environmental parameters or features to compete against)

These stakeholder requirements must be converted into measurable parameters, or environmental parameters for consideration in product design. Commonly used tools are EQFD and EBM.

Significant environmental parameters can be identified by using EQFD [2].

EQFD considers the various stakeholder requirements listed above. It can be implemented in the five steps shown in Table 2.2.

Significant environmental parameters can also be identified using EBM [2]. EBM is a modification of the conventional benchmarking tool that allows the identification of any environmentally weak points of a product in comparison with other products. A relative score ranging from 1 to 5 is assigned to each environmental parameter (very good: 5, good: 4, average: 3, bad: 2, very bad: 1). A gap analysis between the reference product and other products results in the identification of the most significant environmental parameters.

The focus for environmental performance improvement through Ecodesign is the most significant of environmental parameters. In the Ecodesign process, product designers develop performance specifications and functions based on the significant environmental parameters and then seek ways to reduce the impact from those parameters. The product analysis generates the environmental profile of the product.

An "Environmental Profile" is a description of the inputs and outputs (such as materials, emissions, and waste associated with a product throughout its life cycle) deemed environmentally significant. They are expressed in physical quantities that can be measured. Simply put, the ecological profile of a product contains metrics of the significant life cycle based environmental parameters. It provides current information on a product with respect to its impact on the environment. An example is shown in Section 5.3 in the form of a carbon footprint.

Table 2.2 Steps and activities for the implementation of EQFD

Step	Activity
1	Identify the stakeholders requirements and translate them into the environmental voice of the customer (EVOC)
2	Assign a relative weight to EVOC with a scale from 0 to 10
3	Define the relationship matrix between EVOC and environmental parameters with the relationship factor blank (no relationship), 1 (weak relationship), 3 (medium relationship), 9 (strong relationship)
4	Each environmental parameter is multiplied by the relationship factor and summed up for all environmental parameters
5	A relative importance is generated from step 4; from this result, significant environmental parameters are identified

The ecological profile can be used to communicate the environmental performance of a product to the marketplace. There can be a variety of communication tools based on the ecological profile such as environmental product declarations and ecolabelling. Of the various communication tools, the carbon footprint has been considered one of the most appealing in today's market because of society's concern about global warming.

2.3.1.3   Product Carbon Footprint and Management

The "Carbon Footprint" originates from the "Ecological Footprint", although the two differ in concept. The ecological footprint is based on the assumption that every category of energy and material consumption and waste discharge requires the productive or absorptive capacity of a finite area of land or water. By summing up the land requirements for all categories of consumption and waste discharge by a defined population, the total area represents the ecological footprint of that population on the Earth. In short, the ecological footprint measures land area required per person, rather than population per land area [4].

On the other hand, the carbon footprint is not really a footprint as originally defined by the ecological footprint. Rather, the carbon footprint is the carbon profile of a product that identifies the total amount of emissions of greenhouse gases (GHG) including carbon dioxide (CO_2), methane, etc., associated with a product over its entire life cycle. Often the term "Product Carbon Footprint" or PCF is used in lieu of carbon footprint. When data collection is limited to GHG emissions in a LCA, then the totality of the LCA results of the product becomes the carbon footprint of the product or PCF. The carbon footprint, therefore, is a sub-set of the data gathered during the LCA implementation.

There are several approaches in collecting and calculating the GHG emissions of an organization, operational facility, and product leading to the identification of the reduction opportunities and development of a carbon footprint [5]. They are the IPCC Guideline [6], The Green House Gas Protocol [7], ISO 14064 [8], BSI PAS 2050 [10] and ISO/WD 14067 [11]. The first three are for the organization/operational facility and the last two for products.

The IPCC Guideline is an international guide for the calculation of GHG emissions from various emission sources. It sets up criteria for major emission sources in various industries and defines equations for the calculation of GHG emissions from the sources as well as emission factors. The primary use of this guideline is for the development of the GHG emission inventory data of nations, organizations, and operational facilities.

The GHG Protocol offers principles and standardized procedures for the calculation of GHG emissions of an organization [7]. In particular, the GHG Protocol classified GHG emissions based on emission types and sources into scopes 1, 2 and 3. A scope 1 emission is a direct emission from the organization and operational facilities, a scope 2 emission is an indirect emission due to the electricity consumption, and scope 3 emissions include all indirect emissions other than scope 2, including emissions from the supply chain. The calculated GHG emissions thus should be reported in terms of scopes 1, 2 and 3, where scope 3 is an optional element.

The ISO 14064 standard, although international in nature, is less practical in calculating GHG emissions for companies compared with that of the GHG Protocol and less specific in criteria for emission sources compared with that of the IPCC Guideline. Thus, it is not used widely by industry.

BSI PAS 2050 [10], is a guide for the calculation of GHG emissions of a product and service, not organization or operational facility. It is based on the principles and methods of LCA, in particular, ISO 14040 and 14044. It also incorporates in the methods of the GHG Protocol and the IPCC Guidelines. However, it does not provide specific methods and equations for the calculation of GHG emissions. Thus, it is necessary to use the IPCC Guideline method for the development of the GHG emission inventory data from various emission sources.

ISO 14067 working draft [11] is being developed to be an international standard for quantifying carbon footprint of a product and service in its part 1 document, while part 2 is for communication of the quantified carbon footprint to the market. The quantification method is based on LCA, in particular ISO 14040 and 14044, and intends to incorporate salient features of existing GHG emissions methods including all the Guideline, Protocol, and standards discussed above. Since it is in the developmental stage, no definitive methods are available from this working draft yet.

A method for the quantification of the carbon footprint of a product can be proposed in which the salient features of the existing methods are combined. This proposed method is based on two different approaches, i.e. GHG emissions data collection at the organizational level and product level. The methods of the IPCC Guideline and the GHG Protocol were used for the GHG emissions data collection at the operational level. The methods of BSI PAS 2050 [10] and ISO 14067 working draft were used for the GHG emissions data collection over the entire life cycle at the product level. The latter follows the conventional LCA approach in collecting data. Operational and product levels have different degrees of details in the data collection.

Alert

Operational Level:

Normally only scope 1 GHG emissions are collected here. No serious consideration is given to scope 2 and 3 emissions. In depth analysis of the scope 1 process enables the operation and/or organization to use the GHG emissions data for accounting and certification of the GHG emissions.

Product Level:

Collection of the GHG emissions data of a product follows the conventional LCA approach based on life cycle thinking of the product. The degree of detail of the GHG inventory in the manufacturing stage of the product is not detailed enough to be used for the GHG accounting and certification. The degree of details of the GHG inventory data for scopes 1, 2 and 3 is the same.

Table 2.3 GWP of typical GHGs [12]

Species	Chemical formula	GWP_{100}
Carbon dioxide	CO_2	1
Methane	CH_4	25
Nitrous oxide	N_2O	298
Hydrofluoro carbons (HFCs)	–	124–14.800
Sulfur hexafluoride	SF_6	22.800
Perfluoro carbons (PFCs)	–	7.390–12.200

The collection and calculation of the GHG emissions data of a product consists of two parts; product modelling and calculation of the GHG emissions. Product normally includes packaging and the unit of the GHG emissions is kg CO_2 equivalent (eq). The basis for the equivalency is 100-year global warming potential (GWP). Table 2.3 shows GWP values of typical GHGs.

The outcome of the data collection includes: (i) the GHG inventory of the product manufacturing stage including key parts and unit manufacturing processes, and (ii) the opportunities for improvement of the GHG emissions along the supply chain of the product. The identified improvement opportunities will allow carbon management or management of the GHG emissions of the product and its supply chain.

Part 1 Product modelling

A process tree of a product is developed by identifying major processes in the use of raw materials and manufacturing processes of the finished product. The process tree is based on the bill of material (BOM) data and scenarios of the distribution, use and end-of-life stages. Using the process tree, the product system boundary is defined. For data collection, data requirements of the product system, or what data should be collected is defined. Since the product system is vast and includes non trivial processes, it is a normal practice to exclude minor or trivial materials, parts and processes from the system boundary. In this case, fractional contribution of GHG emissions from the parts, materials, and processes to the total GHG emissions of the product can be the cut-off criterion. In addition, it is a normal practice to exclude the following: transport of workers to the factory, transport of consumers to the shop, lighting and heating of factories and offices, and manufacturing and maintenance of capital goods.

For the data collection, the mass flow of all major input and output streams into and from each unit process should be identified and its amount must be quantified. Of particular importance is the data on energy consumption and direct emissions in each process in the process tree. The type of energy that is being consumed or GHG that is being emitted are classified into three categories.

Data requirements if energy source is electricity:

Process level electricity consumption in kWh/kg of main output produced
The emission factor of the electricity is in kg CO_2/kWh

Data requirements if energy source is fuel:
Type of fuels being used (e.g. natural gas, diesel)
Kilogram or liter of fuel consumed and its energy content/kg of main output produced
The emission factor of the fuel is in kg CO_2/kg or liter of fuel
Data requirements if direct gas emissions are being produced:

Type of emissions (e.g. CO, CH_4, N_2O, SF_6, HFCs, perfluoro carbons)
kg of gas emitted/kg of main output produced
The GWP of the gas is the equivalency factor

Data collection of the GHG emissions from each life cycle stage in the product system boundary of a product follows the method described below:

Use of raw materials stage: This stage consists of the acquisition and processing of natural resources, including raw and ancillary materials, and the associated transport of the resources and materials. It is impractical or impossible to collect the GHG emissions directly from this stage; rather, it is normal practice to use the GHG database of materials, processes, transport, and activities. The GHG database is similar to the GHG emission factor, which is the amount of GHG in CO_2-eq/kg of materials or per a certain activity such as incineration of a material.

Manufacturing stage: This stage consists of the manufacturing of parts and the finished product. In both cases, the GHG emissions from the identified emission sources are measured directly or calculated indirectly. The emission sources can be classified into stationary combustion, mobile combustion, process emissions, fugitive emissions, and indirect emission due to electricity consumption [7]. Identification of emission sources involves the identification of the facilities and activities emitting GHG which are included in the product system boundary. Activity data should be collected at the actual point of discharge of the GHG emission. Examples of the activity data include fuel consumption (e.g. Liquified natural gas and Bunker-C oil) for boiler and electricity consumption for motors and mechanical equipments, etc.

Since the manufacturing stage data is from the site, the GHG emission data is the site specific data. This needs to be allocated to the functional unit of the product. Thus, allocation of the site specific data is made to generate product specific data. Conventional allocation rule [27] can be applied for this purpose.

Distribution stage: GHG emissions mainly stem from the fuel combustion of mobile sources. A distribution scenario based on the product weight, mode of transport, transport distance, and number of products transported in one transfer should be made for the calculation of the GHG emissions.

Use stage: A use scenario of the product by the consumer is initially made, including the time of operation in use and the lifetime of the product. The average energy consumption should be estimated by taking into account the product lifetime and consumed electricity (on, off, and standby mode). Since there is no standard for the estimation of the average energy consumption, the scenario used by the Implementing measure of the standby electricity of the Ecodesign directive [13] is recommended.

End-of-life stage: The recycling methods of the recyclable materials and parts and the disposal methods for the remaining waste depend on the actual collection rate and the type and weight of the materials of each part in a product. As such, the end-of-life scenario based on the collection rate, type of materials, and weight of each part in the product should be made for the calculation of the GHG emissions.

Calculation of GHG emissions of the product system

The data collected from each of the product system boundaries above need to be converted into the GHG emissions data. The calculation is simple. You need to multiply the data (e.g. fuel consumption, electricity consumption, the amount of gas emitted, distance transported, disposal method, etc.) by the corresponding GHG emission factors. A complete example is shown in Section 5.3 for a complex set top box.

Emission factors should be collected from the site investigation, if possible. When emission factors from the site investigation are not available, then national emission factors should be used. When national emission factors are not available, then emission factors given by IPCC can be used.

In many cases, the GHG emission factors of materials, processes, transport, and activities are not available from the IPCC Guideline. Thus there is a need to calculate the GHG emission factors by converting the LCI database of materials, processes, transport, and activities into the corresponding ones. Here, characterization of the LCI database into the global warming impact category using the GWP 100 years can generate the GHG emission factors. For instance, converting the LCI database of polypropylene, cardboard, compounding extrusion, 11.5 t truck transport at 60 km/h speed, Korean electricity, and plastics recycling into the GHG emission factors via characterization gives 1.37 kg CO_2-eq, 32.6 kg CO_2-eq/kg, 0.24 kg CO_2-eq/kg, 29.1 kg CO_2-eq/t-km, 0.42 kg CO_2-eq/kWh, and −0.96 kg CO_2-eq/kg, respectively [14].

For the calculation, it is better to divide the calculation into two separate steps: one is at the process level and the other at the product level.

For each process:

1. Arrange collected data in three categories: fuel, electricity, and direct gas emissions, either in kg liter of fuel, kWh electricity, or kg of gas/kg main output produced.
2. Convert the data by multiplying relevant emission factors for the energy sources used, or the GWP emission factors so that the data can be converted into CO_2 equivalent GHG emissions/kg main output produced.
3. After allocation, the GHG emissions per main output from the unit process to be used for the finished product can be obtained through mass balance. (If no allocation is necessary, the result from step 2 is used in the next step).

For the product system:

Add the GHG emissions data from all individual unit processes to obtain overall GHG emissions of the product system. Note that the contribution by individual unit processes to the finished product should be accounted for in this summation (e.g. this is the same as the Life Cycle Inventory (LCI) calculation procedure).

The GHG emission of a product can be calculated by adding all the GHG emissions from all life cycle stages as shown in equation below.

$$GE = GE_{Raw} + GE_{Mfg} + GE_{Dist} + GE_{Use} + GE_{Eol},$$

where

GE: GHG emission of the product (kg CO_2-eq)

GE_{Raw}: GHG emissions from the use of raw materials stage (kg CO_2-eq)

GE_{Mfg}: GHG emissions from the manufacturing stage (kg CO_2-eq)

GE_{Dist}: GHG emission from the GHG emissions of the distribution stage (kg CO_2-eq)

GE_{Use}: GHG emission from the GHG emissions of the use stage (kg CO_2-eq)

GE_{Eol}: GHG emission from the end of life stage (kg CO_2-eq)

Alert

Product carbon footprint (PCF)

International standardization of the PCF is underway in ISO/TC 207. The standard consists of two parts: quantification of the carbon footprint of a product and communication of the carbon footprint to the market. The fundamental principles and methods of LCA in ISO 14040 series and the type III environmental declaration in ISO 14025 have been adopted for the quantification part and the communication of the PCF standard, respectively. This implies that SMEs may find it difficult to comply with the PCF standard. The requirements on data collection and setting product category rules for communication may be too burdensome and expensive. Industry may need to find a less demanding route for the declaration of PCF once the standard becomes available in the market.

2.3.2 Resource Use Efficiency Aspects

The analysis of resource use efficiency considers the possibility, as well as the ease of recycling of a product when its useful life (i.e. value lifetime) ends. In this analysis, the recyclability of the product is often evaluated. To analyze the ease of recycling, several factors, such as the use of composite materials, number of plastics used, coating and painting of the external plastics, are evaluated.

In general, recycling and reuse are considered important design factors for products such as photocopiers. Recycling and reuse considerations incorporate ease of recycling and reuse of parts in the product's structural design. This practice is rather common for those products commonly leased. Although the reused parts from any leasing business have to have the same quality as new products manufacturers of energy-related products intended for sale only have to prove to the public that a detailed quality analysis had been applied to the reused parts (cf. Section 2.4).

Design for easy recycling can be achieved in two consecutive steps. First, calculate the recyclable rate of the product and then set up the recycling rate target of the product being developed. Second, develop solutions to achieve established targets.

In the first step, the recyclability rate can be calculated by analyzing own and competing products in terms of material identification of the parts and measuring their weight based on the recyclability calculation formula:

$$\% \text{ Recyclability} = (\text{total weight of all recyclable parts/product weight}) \times 100$$

Establish the recyclability target based on the regulatory information on recycling from various countries including the EU's WEEE regulations and the calculated recyclability rate of your own product. In the event that competitors' products have better performance and features with respect to material selection, disassembly, product structural design, and ease of recycling, those aspects should be reflected in the redesign of the product.

In the second step, the following principles can be applied when determining solutions for product development, including packaging, which facilitate recycling:

– Minimize the number of plastics used. Mono-type materials are the best.
– Avoid painting and surface treatment of the external plastics.
– Avoid structural design that makes disassembly difficult when different materials are used, such as metal inserts in the plastic parts.
– Identify material types when using plastic parts.
– Select recyclable materials.
– Implement structural designs that enable easy and quick disassembly of the parts requiring frequent service, the parts with higher recycling value, and those containing hazardous substances.
– Minimize the number of joints such as bolts and nuts, and ensure that disassembly can be done through the use of regular and readily available tools.

Once solutions are developed, resource use efficiency can be evaluated using a matrix similar to the one shown in Table 2.4.

For each evaluation criteria, there are corresponding evaluation scores ranging between 0 and 3. In addition, a weight (representing relative importance or priority) is assigned to each evaluation sub category. The weighted evaluation score of each sub category can be obtained by multiplying the evaluation score by the weight. Based on the weighted evaluation score, the resource use efficiency of the product

Table 2.4 Evaluation of product resource use efficiency

Category	Sub category	Evaluation criteria	Target	Evaluation results
Product weight	Product weight	Measured weight of product, accessories, and batteries	Weight of product	Product weight, Battery weight
Recyclability	Recyclability	% Recyclability = (total weight of all recyclable parts/ product weight)×100	Large product: 75%, IT product: 65% Small product: 50% (cf. also WEEE targets![a])	% Recyclability
Easiness of recycling	Coating and painting of external plastics	% coating/painting not applied = (1−number of parts with coating and painting/number of external plastic parts) × 100	Yes or No per product	Number of parts with coating and painting
	Use of composite materials	Check the presence of composite materials used in the plastic parts exceeding 20 g weight	Yes or No per product	Parts names concerned and type of composite materials
	Number of plastics used	Plastics in the external cover and housing	Number of plastic types	Number and type
	Material name of the plastic parts identified	% identification = (number of parts identified/total number of parts considered) × 100	100% material identification	Number of parts considered Number of parts with material ID

[a]Numbers in this table are for illustrative purposes only

being developed can be calculated. If the weighted score does not meet the target, iteration may be necessary to choose an alternative design.

2.3.3 Degree of Environmental Hazard Aspects

The analysis of the degree of environmental hazard considers critical substances and toxicity of materials and parts used in the product (including packaging). The analysis includes screening of parts from suppliers with respect to the presence of hazardous substances regulated by law and the company's own requirements. Controlled hazardous substances contained in the supplied parts and product include:

– Class I controlled materials (all six RoHS regulated substances)
– Class II controlled materials
– Company specific controlled materials
– Halogen containing materials
– Volatile organic compounds (VOC) emissions during the use of the product

The analysis can lead to the substitution of materials and/or parts going into the product.

When analyzing the degree of environmental hazard of a product, and developing solutions to meet the requirements of the regulations, the relevant hazardous material related requirements can be identified by reviewing the regulations of various countries, and other requirements related to ecolabelling programs, environmental non-governmental organizations (NGOs), and competitor trends. This should include the identification of requirements to be reflected in the product design, and whether or not to pursue ecolabel certification. In particular, the following should be considered:

– Meet the legal requirements from the EU, China, Japan, Korea, and other countries on the RoHS-like regulations.
– Meet the ecolabelling requirements of a specific product group with respect to hazardous substances.
– Prepare for expected regulations on substances such as polyvinyl chloride (PVC), antimony, beryllium, and brominated flame retardants (BFRs) by listing them as non-use, hazardous, or controlled substances.
– Select bioplastics and materials that reduce GHG emissions as part of a competitive differentiation strategy.

Of particular significance is the implementation of a parts control system to ensure the absence of newly established non-use substances in components and parts. A control system should typically require approval of the parts after (a) independent verification (perhaps requiring laboratory analysis) and include (b) control of the parts' manufacturing system to ensure the same parts are made as those passing any verification or the laboratory test.

New materials, using no PVCs or antimony, should be developed in co-operation with material suppliers. The entire control process can be implemented by adding the newly enlisted non-use substances into the existing parts control process such as the Green Partner or Eco-Partner program. Don't forget to avoid the "old" very toxic and, in many countries, already prohibited materials like asbestos, PCB, etc. also in those countries where the application is still allowed! These substances are usually no longer used in consumer products but in plants and in projects.

Although the control limit of hazardous substances may be met at the parts level, the product level should also be verified to ensure an absence of hazardous substances. Factors such as heat dissipation from the product during the use stage may trigger the emission of VOCs, which could thereby exceed the regulated level.

The degree of environmental hazard of the product can be evaluated using a matrix similar to the one shown in Table 2.5 which gives an example of the evaluation of the environmental hazardousness of the product including the controlled substances, halogen free parts, and VOC emissions.

The same evaluation logic as that described under Table 2.4 can be applied. Based on the weighted score, the degree of environmental hazard of the product being developed is evaluated. If the weighted score does not meet the target, iteration may be necessary to choose alternative design options like, for example, the selection of alternative parts.

2.3.4 Extended Material Aspects

The analysis of the extended use of materials should include consideration of the use of recycled materials or reused parts and components and the use of renewable materials in a product. For the assessment of the use of recycled materials, use of recyclable materials and reusable parts of the waste product are considered. In the case of the use of renewable materials, selection of plant derived materials can be considered.

The extended material aspects of a product are normally assessed using an approach described in Table 2.6.

The same evaluation logic applies here as that described under Table 2.4. Based on the weighted score, the extended material aspects of the product being developed can be evaluated. If the weighted score does not meet the target, iteration may be necessary to choose an alternative design option, such as the selection of alternative materials and parts.

Table 2.5 Evaluation of the environmental hazardousness of the product

Category	Sub-category	Evaluation criteria	Target	Evaluation results
Class I controlled substances	Six substances under the RoHS regulations	Check the use of the six regulated substances in the product, packaging, and batteries	Within the control limit of the six RoHS-regulated substances	Meet the regulated (target) value? For product: avoid all six substances Packaging and battery: four heavy metals
Class II controlled substances	Specific substances controlled by the company	Check the use of class II controlled substances in the product, packaging, and batteries	Within control limit of the company-regulated substances	Meet the company-regulated (target) value?
Halogen-free	Power and other cords	Check the use of halogen-free materials and no PVC use	Variable per each product group	Power cords with no PVC use halogen-free (power) cords
	Structural parts	Check the use of halogen, free materials especially no PVC/Tetra-bromo-bisphenol A (TBBPA) use	Yes, if no TBBPA use Variable per each product group	No PVC use No TBBPA use All halogen-free
	Printed Wiring Boards	Check the use of halogen-free materials and especially no TBBPA use	Yes, if no TBBPA use Variable per each product group	No TBBPA use All halogen-free
	Packaging	Check the use of halogen-free materials	Yes, if no PVC use	No PVC use All halogen-free
VOC	VOC	Check the compliance with the VOC emission requirements in the ecolabelling and European Computer Manufacturing Association (ECMA)-328 criteria	Variable per each product group	Measure VOC level (dust, ozone, benzene, total VOC, toluene, etc.)

Table 2.6 Extended material aspects of products

Category	Sub category	Evaluation criteria	Target	Evaluation results
Use of recycled materials	Use of recycled materials	Amount of recycled materials present in a product	Variable per each product group	Yes/No of the use of recycled materials Part name, type of recycled material, percentage recycled content in the product
	Use of reusable parts	Amount of reused parts and components in a product	Variable per each product group	Yes/No of the use of reused parts Part name; percentage reused parts and component content in the product
Use of eco-friendly materials	Use of plant derived materials	Used for products, packaging, and accessories, except paper packaging	Variable per each product group	Yes/No of the use of plant-derived material Part name, type of material, percentage plant-derived material in the part
	Use of more eco-friendly materials (excluding plant derived materials)	Used for products, packaging and accessories, except plant derived plastics	Variable per each product group	Yes/No of the use of eco-friendly material Part name, type of material, percentage eco innovation material in the part (these are parameters for the evaluation of the eco innovation aspects or extended material aspects: same as above)

Checklist for product analysis

Assessment questions	Answer	Comments	Recommended follow-up activities	Related sections
Have you analyzed the environmental aspects of the product from the perspective of the product's life cycle?	Yes ☺ No ☹	The intent of the product's environmental aspect analysis is to identify the significant life cycle based environmental parameters of the product for improvement through Ecodesign	Perform analysis of the product's environmental aspects using tools such as LCT matrix or LCA	2.5, 3.3

(continued)

Checklist for product analysis (continued)

Assessment questions	Answer	Comments	Recommended follow-up activities	Related sections
Have you analyzed the environmental aspects of the product from the perspective of stakeholder requirements?	Yes ☺ No ☹	The purpose of the stakeholder requirements analysis is to identify the significant environmental parameters relating to stakeholders in order to improve the product through Ecodesign	Perform analysis of the stakeholder requirements using tools such as EQFD and EBM	3.3
Do you intend to develop and communicate the product carbon footprint (PCF)?	Yes ☺ No ☹	The product carbon footprint is an environmental profile of the product	Follow international standards, such as ISO's PCF standard	
Have you evaluated the resource use efficiency of the product?	Yes ☺ No ☹	The resource use efficiency analysis considers the possibilities of recycling as well as ease of recycling of a product when its useful life ends	Evaluate the resource use efficiency of the product – calculate recycling rates	2.5, 3.3
Have you evaluated the degree of environmental hazard of the product?	Yes ☺ No ☹	Consider critical substances and toxicity of materials and parts used in the product (including packaging)	Evaluate the degree of environmental hazard of the product. Comply with RoHS, etc. Define your own list of class II controlled substances	2.4, 2.5
Have you evaluated the extended material aspects of the product?	Yes ☺ No ☹	Reuse of parts and components, using recycled or even renewable materials, are aspects of innovation and cost efficiency	Evaluate rate of recycled material, reused part, components and eco-friendly material present in the product	2.5

2.4 Production Analysis

In this section, the analysis of the environmental aspects related to processing materials and substances is described. The view of these materials and components is from the perspective of production, development, and purchasing but also includes potential risks associated with the suppliers. Reuseability and recycling of components and materials can be investigated through the disassembly and reuse analysis.

The selection of materials is also important for the technologies applied in the production processes and for the planned recycling. An analysis of these processes should be undertaken to identify potential improvements or material substitutions.

2.4.1 Information on Materials and Substances

In a globalized world, many producers have become assemblers. They purchase most of the component parts within their product's value chain. As a result, the manufacturer, or assembler, will most often have little or no information about the environmental aspects or composition of the components.

In order to conduct a competent inventory analysis of its own production to determine environmental impact, the company needs to know which materials are supplied directly or in components. In addition, customers, governments, NGOs, or regulatory agencies may require information about the use or application of prohibited hazardous substances. Also, some of this data is relevant in order to ensure workplace safety. Threshold limits in the component or complete product must be calculated for some substances, and this is only possible if the other quantities in the component are known.

There is also some potential risk of price increases resulting from material scarcity or market speculation. Some prices have increased many times over and, for some substances like indium, significant price increases should not be surprising. Substances such as indium should be evaluated to determine price risks. For those materials with high price risk: (a) substitute materials should be sought, (b) quantities consumed significantly reduced, or (c) a recycling process developed. For other material such as noble metals, the environmental impact through mining is extremely high.

During the analysis, it should be determined whether a complete production information system is available. Such a system should facilitate finding answers to questions about the material record of all chemicals or materials used, especially hazardous substances. Rare and risky substances should, therefore, be easily identified. All volumes of materials used in the production of the products should be available. The procurement system should require the supply of the relevant information. A system should guarantee the correct supply of materials and components with the desired environmental properties. Regular reporting from suppliers must be ensured.

2.4.1.1 Special Information About Materials and Substances

Legal requirements, e.g. from REACH [15] demand that information about substances in down-stream use be available, i.e. in applications. As about 30,000 substances are currently under discussion, it will not make sense to restrict the information of the contents of a product. Therefore, the total content of substances should be known to a significant degree.

Lists of substances to be avoided can be provided by associations like the DigitalEurope (voice of European digital economy including IT, communication and consumer products) [16] or the German Electrical and Electronic Manufacturers' Association (ZVEI) [17]. These lists contain those substances under discussion that will probably be restricted in the next few years. These lists can also be used in risk analysis, and to identify where alternative materials are likely to be needed.

Recycling information about hazardous substances in waste products is not currently widely required by recyclers. Most recyclers have experience with the usual E&E (electrical and electronic) products. Nevertheless, it might be beneficial to know which substances have residual value.

In this complicated area, standardization is necessary (e.g. IEC PAS 61906) and already under development (e.g. IEC 62474). Substance information is offered on three levels:

Product – e.g. for recycling costs
Component – e.g. for reuse
Material – e.g. for evaluation of required threshold limits, cost development calculation.

These standards also define the data format, thereby allowing software companies to offer programs for many users. It will not make sense for every company to develop its own program, as data will have to be freely exchanged.

To reduce the amount of data by factors of ten to one hundred, different associations and companies have developed standardized average material records for component families [18]. In addition, the concept can also be extended to complete standard components and products which are often very similar. Equally, information is available for plastics, buildings, and/or cars [19]. Information about substances is also a necessary input for LCA. Data sets are already available for buildings, cars, and plastics from the various organizations. In the near future, it can be expected that suppliers will be required to provide this kind of data with every order. Additionally, standardized procedures for material declaration will be required and will, most likely, soon be available.

Materials information or component information must fit the production information system. From there, an input/output analysis is undertaken. The material record of the complete product will also require this information. Environmental data for production and for components or products are different. Many materials used in production will never end up in the product. Naturally, toxic effects can vary from person to person and, thus, legal requirements can also differ. In addition, information to the public may have different requirements. Thus environmental improvement of the product and of production should be planned individually.

The management of materials in products and components, whether hazardous or not, should be part of the management system [cf. Section 2.5]. Environmental information about standard plants and especially energy consumption, is often not directly available from producers. With motors or pump controls, control of ovens, fans, power supplies, etc., energy consumption can be reduced by about 50% compared with the standard configuration [20].

In some regions, asbestos and other very toxic materials are still allowed. Regardless, information on all toxic materials is needed because legal requirements in one country are not lessened because of less restrictive regulations in other countries. Insurance companies may well choose to deny insurance coverage if it were discovered that such toxic materials were used! And if illnesses were to occur, there could be successful lawsuits.

Purchasing departments can, as a first step, use the above mentioned lists from industry associations for substances already prohibited in many countries, and develop additional lists of substances to be avoided for their own business.

When examining energy consumption of a production line, the most energy consuming components are usually drives, pumps, ovens, and exhaust systems. Often, only one of these consumes the largest percentage of the total energy. With improvement of this main component alone, in some cases, up to half of the energy consumption of the complete production line can be saved. It is often enough to build in a control element to reduce energy consumption. However, in most cases a new main component, such as a more controlled soldering oven, would be needed. Planning for this kind of component change may be lengthy, depending on whether the oven supplier (in this example) actually has such an energy saving component available. Nonetheless, the identification of the energy saving opportunity may well stimulate the oven manufacturer to develop the more energy efficient component.

Management usually has to authorize this kind of equipment replacement because of the level of investment involved. Many projects are rejected on the argument that the payback period on the investment is too long, although, payback times can be rather short if the frame of reference is the entire life cycle of the product or the whole production process.

The afore-mentioned subjects should be integrated into the order's management process and should be part of the overall energy management of a factory or plant. To receive environmentally improved components from suppliers on a timely basis, the following steps are suggested:

1. Plan ahead for energy saving investments.
2. Calculate payback times within the life cycle of the product or the whole production process, and include energy price trends.
3. Involve potential suppliers early in the process and certainly before any orders are placed.

Check which additional special data your company requires for the business. Define special materials with risks in product or production. Also identify components in the production line with high energy consumption.

Alert

Product vs. production information:

Be careful not to confuse product information with production information. Materials used in a production process often do not remain in the product. The legal requirements for substances in products are normally different from those used during production. Also, the internal and external target groups that might be interested in the information will be different.

Alert

Data becomes outdated:

Data supplied from authorized institutions or industry associations can often be old or out of date. Especially LCA data might therefore not be comparable.

2.4.1.2   Disassembly Analysis of Components

For car manufacturers, disassembly information is part of the first conformity assessment. Using the standardized "Recycling evaluation method" (REM) [21], a theoretical value for recyclability can be calculated. For the E&E industry, with its many product variations and modifications, this method is not applicable. But every manager should be aware of the degree of ease with which his product can be disassembled. Such information can provide valuable clues for assembly, and reduces assembly costs. Also, those companies who choose to reuse components already have to plan for easy disassembly. Software programs are available with standard disassembly times [22].

Alternatively, a recycler can measure the disassembly time of a product and provide important feedback for those joints that are difficult to separate. Difficult disassembly is a synonym for higher costs! Thus, the design and the hierarchical structure of the product and components should be modified in cooperation with the suppliers.

In summary, disassembly analysis can help to significantly reduce assembly costs! This analysis, together with planning for reuse of components and designing for minimum joints and junctions, facilitates the reuse of components. If components should not be reused, the analysis can be different. For the recovery of pure mono-type plastics, disassembly is not the only way. It can also be done by special shredding technologies followed by automated sorting.

2.4.1.3   Reuse Analysis of Components

Reuse analysis is more than disassembly analysis and can, for SMEs, usually only happen in cooperation with suppliers. Reuse is an objective of product planning together with ease of disassembly (e.g. by avoiding joints that cannot easily be taken apart). Reuse analysis also includes the calculation of costs for requalification of the components.

Reuse of components and products offers the highest value-added, especially for capital goods. Components from these products or plants are often like new (i.e. undamaged), and in fact frequently cannot be distinguished from the new ones. The lifetime of the product or component can be long, often up to 30–50 years. Customers are known and components can be taken back if the customer plans to buy a new product.

Reuse analysis can be also done for service applications and as input to new product development. The highest value-added is when new products are manufactured with components that are "qualified-as-good-as-new" (e.g. defined in IEC 62309).

Standards are in place that define the conditions for requalification from the take-back process and include the documentation and customer information requirements [23].

The reuse analysis consists of:

- A determination of valuable components
- The selection of the qualification procedures for the components (testability)
- A calculation of costs and benefits for the refurbishment process
- An investigation into the possibility of planning for several product generations
- A check of the acceptability of the requalified components by customers

Example

The analysis of the mounting plate of a computer showed that all plates were different by some mm. The cost of a new plate was in the range of 20 €. As no stress occurs on the part, visual testing is enough to investigate the state of this component. After some cleaning, it can normally be reused as is. Only this simple part constitutes 3–4% of the total costs. Therefore, if this part is qualified as-good-as new and reused, the product could become cheaper by about the same percentage.

For a cost/benefit analysis of a component, the costs and benefits should be compared, as shown in Table 2.7.

Typical costs for testing, disassembly, and cleaning can be in the range of 1–2 €. The value added for some parts can also be very high especially for components "qualified-as-good-as-new" according to IEC 62309. It should be mentioned that a modern quality management system should be installed together with a complete and documented procedure for quality testing. No additional risk should be transferred to a customer and customers must be fully informed about the reuse. As a benefit to the customers, the manufacturers, and the environment, the new products with re-qualified components can be much cheaper and of better quality than new ones.

Table 2.7 Benefits and costs for the refurbishment and reuse of components

Benefit	Cost
– Avoided cost for a new component (price decrease of component over time has to be considered)	– Quality testing (e.g. acc. to IEC 62309)
	– Disassembly
	– Cleaning
– Sales of some parts are also possible as spare parts, for service and possibility of other applications	– Software testing and upgrade
	– Recycling/deposition cost
	– Documentation and information
– Recycling benefit for valuable material	

A useful strategy is to evaluate all components "of interest" when designing a new product. Those products that have components that are planned for reuse should be designed in such a way as to ensure easy removal. In addition, the application of such components should be planned for several product generations, because the first totally new products will only be returned after some years and the maximum take-back might only occur when the next generation of the product is placed on the market. The full potential of the used components could, therefore, only be realized when the next product generations are produced. The percentage of reusable components can typically be between 15–25% of a complete product. Some components can have several lives.

For public tenders for large projects such as train stations, metro systems, or power plants, questions about the lifetime of components can arise and can become a determining factor for the customer's cost planning. Examples can be found in the cases of Xerox or Siemens Healthcare [24].

Alert

The highest benefit from the re-qualified components can be achieved if the product is produced as a serial product and the components can be integrated into the production line of a new product. But

- Requalification can also be done by suppliers who have the required test equipment
- SMEs can organize take back of their used equipment, perhaps together with a competitor to achieve economically viable volumes
- Electronic components are often difficult to test and can therefore often not be reused in new products
- The highest value-added is often located in simple and usually inexpensive parts
- Reuse is also possible for components from the waste stream of short-life products
- Additional advantages are found with low age products because components may be like new
- Leasing is not the only way to get products back for reuse. In many cases, a good customer relationship can be built which enables reuse

After it has been decided how to manage the reuse of components, an investigation of the recycler market will show variation in qualifications and expertise. Some can only shred, others can disassemble and supply valuable components, some can provide storage, some have a license for special waste treatment, and for others it is possible to resell plastics with quality as new.

2.4.1.4   Organising Recycling

The knowledge about the way a recycler operates and knowledge about his technical equipment is essential to select appropriate and useable recyclable materials. Some recyclers only operate shredders; others disassemble components and housings and sort plastics. A few can also granulate plastics with "as new" quality. Many companies also forget that the recycler is also a supplier. For some, the recycler seems only to exist as a black box. Therefore development engineers, who follow a lot of guidelines for Ecodesign, become quite astonished to find that not all recyclers can recycle material that has been selected as most environmentally compatible! To optimize recycling of their own products, an analysis of the recyclers and their recycling technologies is unavoidable. If no reuse of components is planned, and no materials are to be reused, the recycler with a simple shredder might be the best choice. But if a company plans to utilize expensive technical plastics, such as polycarbonates, and wants to reuse them, then either (a) a chain of several recyclers that collectively can disassemble, sort, collect to achieve amounts sufficient for regranulation, or (b) a special plastic recycler will be required.

So the first question should be: How is the product collected for recycling? Are there legal requirements? In Europe the WEEE enables collection and/or individual take back. It also requires the take back and recycling of some capital goods like medical equipment. Usually, recyclers are very experienced with the standard E&E products and usually know how to dismantle them.

One group of recyclers may prefer a detailed disassembly because they can earn more money with clean fractions. Another group may recycle by using shredder technology. Both approaches enable separation of special materials as required by European law. A company in the recycling business can compare component or commodity prices with the cost of operating the different options for recycling. Some recyclers already offer the disassembly of spare parts and the storage of these parts for service purposes. Combinations of technologies can include thermal treatments with heat recovery or chemical conversions. They use or mix plastic fractions. Some need a high temperature furnace to, for example, crack the brominated flame retardants [Annex II of WEEE].

The ranking of reuse of products and components after take-back has become a higher priority for the European waste legislation. So it will be worthwhile for every company to think about reuse and recovery together and not concentrate only on recovery. In addition, recovery rates will have to integrate the quantity of reused components and products. As it is planned to increase the required recovery ratio, reuse becomes additionally important because its ratio can be included.

If a recycler requires information about any disturbing impurities, a special checklist can be used, such as the one developed by the European Association for Household Appliances (CECED, French: Conseil Européen de la construction d'appareils Domestiques) or the European Recycler Association(ERA). On the other hand, a company can decide which materials should be separated, and can offer these materials if they are easy to dismantle (e.g. plastic housings). Generally, the total cost will decide what is selected. In some cases, a company's own system can be of interest, such as the system for printer cartridges or special recycling of CDs for expensive polycarbonates. Such decisions can usually only be taken by big companies alone or can become a collective solution. However, generally the collected quantities are too small for special treatment. Internationally, a take-back standard is being discussed by countries where no legal requirements currently exist.

The quality check conducted by the recycler should cover compliance with the legal requirements to ensure a high standard of recycling. In many countries, recyclers have to be licensed by the government, and their recycling processes have to be certified by auditors. If the quality of the recycler does not meet the requirements of your company, any co-operation with the recycler should end.

It must be mentioned that the responsibility for recycling of a company's own products in Europe is a regulatory requirement for manufacturers.
The related investigation should include questions such as:

– How is my product recycled? What technologies are used?
– How can my company profit (for example, with the selection and recycling of special materials or cost reduction with the avoidance of some materials)?
– How can we participate in a system or should we do it individually? Together with competitors?
– Does a company make a profit or is recycling a cost factor? If it is expensive, how can it become cheaper or profitable?
– How can a company avoid or reduce recycling costs? Actions should be taken together with industry associations, competitors, government or individually.
– Which approach fits best to all markets of the product worldwide?
– How reliable are the recyclers (legal requirements)? Is there a risk for the company?

This recycling and recovery analysis will help you find the best technology for your product. In conclusion, environmentally poor production processes should be changed.

2.4.2   Production Process Analysis

According to Graedel & Allenby [25], three classes of processes/systems can be distinguished:

- Type III is cyclic and sustainable
- Type II could be approached today by developing feedback and cycling loops (especially for scarce materials)
- Type I loses resources.

Everybody can evaluate in which class his process is located: Today most will certainly be the type I process.

The simple matrix concept from Graedel & Allenby yields a more complete picture of a production process. It helps to evaluate the sustainability of the process and goes beyond LCA. The following items could be evaluated for the different life cycle stages

- Process compatibility
- Materials compatibility
- Component compatibility
- Performance
- Energy consumption
- Availability
- Cost
- Competitive implications
- Application system (use by different customers for different purpose)
- Others

Evaluation can occur in four steps, from "no concern", "minor concern", "moderate concern" to "significant concern", corresponding to the numbers 1–4, respectively. A group of experts from a company can do the process evaluation themselves, and should develop if necessary its own checklists. An evaluation of different process options afterwards could happen using the following criteria:

- Toxicity/exposure
- Environmental effects and impact, also by life cycle inventory
- Production criteria (investment, easiness, etc.)
- Social, political, and legal trends
- Cost and technical trends

The same ranking method could be used as for the process audit above. The most important point must be to get a decision to change the process in those areas where there is significant concern. Perhaps a more precise analysis will be necessary.

Nevertheless, a trend analysis has to follow these evaluations. If the technology already poses hazards to the environment, to health, or creates any kind of potential risk, then it should be changed or a new technology be developed if an alternative is not available.

It is an important lesson – substances that are going to be prohibited in one country will probably be also prohibited in other countries. Substances with high environmental impact, or rare (or scarce) substances will be restricted sooner or later, or become too expensive. Therefore, when information about any forthcoming legislation is released, one can expect similar regulatory action in other countries. Even if processes that use already targeted toxic substances are run in a closed loop, the time for the process may well soon run out.

Many alternative processes may exist for the same task. Soldering to form fixed joints can in many cases be replaced by compression technology without using chemicals. Preventive health check costs for employees dealing with very toxic lead can be saved, no lead emissions occur and many more problems combined with the application of lead can be avoided. Other solder technologies are also available or adhesives can be applied. An environmental evaluation and a cost comparison of the different technologies will probably show interesting opportunities.

Alert

Managers may choose to switch from a toxic substance to a new substance estimated to be non-toxic. Unfortunately, there may be little known about the new substances, and thus risks are created. When substances such as chloro carbons, chlorofluoro carbons (CFC), and some flame retardants were introduced, the environmental or toxic effects were only known years later, and then substitutes were actively sought out. This is inefficient. In the future, more information will be required by legislation (e.g. in REACH regulation). In many cases, chemicals may not be necessary at all.

Checklist for production analysis

Assessment questions	Answer	Comments	Recommended follow-up activities	Related sections
Is there a systematic procedure installed to gather all required information about chemicals, materials?	Yes ☺ No ☹	Integrate the software systems of the suppliers, use checklists, define responsibilities, also gather quality and environmental information	Optimize system	2.3, 3.3, 3.4, 4.3, 4.4, 4.5
Is it certain that all components or materials are supplied with the correct environmental properties?	Yes ☺ No ☹	Random checks necessary	Optimize system	2.3, 3.3, 3.4
Is it checked by a disassembly analysis which components form a barrier for reuse?	Yes ☺ No ☹	Simple connections required, otherwise reuse too expensive	Analyze all relevant components	2.3, 3.3

(continued)

Checklist for product analysis (continued)

Assessment questions	Answer	Comments	Recommended follow-up activities	Related sections
Is assembly improved by disassembly investigations?	Yes ☺ No ☹	Easy disassembly often means easy assembly! High cost reduction in production possible	Apply software	2.3, 3.3, 3.4
Is reuse evaluated and is a systematic approach for reuse of components and products in place?	Yes ☺ No ☹	Long-term planning required, high cost reduction possible	Install quality system according to IEC 62309	2.3, 3.3, 3.4
Does market intelligence identify which material can easily be recycled?	Yes ☺ No ☹	Targets: Clean fractions, monotype plastics (material recycling is easier with capital goods as producer gets back his own product)	Analyse markets e.g. with plastics Install own system	2.3, 3.3, 3.4
Do recyclers confirm which recovery procedures are best for the product and environment?	Yes ☺ No ☹	Possibility and reality is often different with recyclers	Observe market situation	2.3, 3.3, 3.4
Is there a production process evaluation in place considering environmental impact?	Yes ☺ No ☹	Changes have to be planned in time. Otherwise jobs are at stake, competitors could be quicker	Continuous evaluation	3.4, 4.4
Are trends determined for production processes and measures for change?	Yes ☺ No ☹	Observe forthcoming legislation	Ongoing	2.5, 3.4, 4.4

2.5 Management Analysis

2.5.1 Current Situation

In general, an organization should check which management systems are already available (cf. Section 2.1). Often there are quality management, environmental management, and health and safety management systems available. Many companies have opted for an integrated management system that combines all of them.

Ecodesign, as a process, can be integrated in ISO 14001 as well as in ISO 9001. Related requirements are listed in Annexes IV and V of the Ecodesign directive. It should be analyzed in detail to determine which environmental processes are already installed besides the normal processes in industrial environmental protection (cf. Section 2.1). The following processes are usually either not in place or are incompletely installed in companies and therefore some systematic approaches are missing:

- *Chemicals and materials*: Purchasing, development, treatment and storage, production, hazardous substances, phase out of hazardous substances, release of hazardous substances, availability of required data. Reuse and resale of materials. Is production software available that deals with the flow of materials? Are required data directly provided by the suppliers?
- *Energy*: Analysis of consumption, methodology of measurement, and measures for reduction.
- *Take back, recycling and recovery*: Collection organized (difference between consumer and capital goods), recyclers evaluated and selected, customers informed.
- *Reuse*: Components for reuse analyzed, check for reuse (also as new) in own production, refurbishment of whole products, resale of components and products, application of reusable components as spare parts.
- *Trends*: Legal trends, new standards, competition, comparison with own strategy.
- *Risks*: Cost evaluation, potential time schedule, measure for development and production.
- *Innovation*: Systematic search for alternatives with high reduction potential for materials and processes, evaluation, introduction of innovation process.
- *Production technology evaluation*: Evaluation for sustainability.
- *Communication*: Success to public and customer, crisis communication, preparation of information documents such as environmental product declarations according to ISO 14025.

In all cases, management procedures are required. All management systems nationally installed should also be in place at international locations. Management approaches also require the systematic application of tools. In a kind of toolbox, these tools could be offered by an internal department or by consultants.

Examples include: Design rules, LCA, environmental product declaration, EQFD, emission trading, and PCF. Whereas design procedures are standardized for products as shown in IEC 62430 [26], plants or software recommendations are lacking. In Annexes 1 and 2 there are examples for both.

Reporting structures are another part of a good management system. Aside from legally required industrial environmental reporting, more and more data are required for products and production, including quantities of materials, hazardous substances, quantity of green products in portfolio. Often, the reporting of the environmental costs can improve management awareness. In any case, a report to the board member responsible for environmental issues, which includes improvement targets and measures, should be provided annually. A review of the achievements and outcomes is also necessary.

In Annex 3, a simple scheme for internal environmental cost reporting comparable to quality cost reporting is proposed. Environmental reporting in more detail should be a part of the annual business report or be a part of an additional sustainability report. Minimum requirements for such reports are often set by the government.

2.5.2 Responsibilities

When environmental problems occur and the justice system is engaged, responsibility is usually assigned by a court. To avoid such problems in advance, a clear structure of responsibilities from the board down to the environmental processes should be organized. In designating accountability, everyone is informed about his respective responsibilities. In complex company structures, a level concept can be applied, as shown in Table 2.8.

In some cases, all of the related tasks and responsibilities are described in a booklet about the "Basic Principles of Environmental Protection" authorized by the responsible board member. Such a booklet can also be given to interested parties outside the company. Such a guide is beyond the detailed description of the processes and responsibilities in the process handbook in a management system.

Employees participating in associations, government, or standardization committees with environmental tasks, should also receive regular updates about related trends.

Alert:

A systematic organization does not require full-time experts. Some managers may have the groundless fear that more employees are needed for such an organization.

Also, the process for designating responsibility does not require much work, but does enable a systematic structure. Wherever responsibilities for Ecodesign are "naturally" with a special manager, frequently the product manager, a letter to him or her with the information about the environmental duties might be sufficient.

Table 2.8 Responsibility levels in a company with responsible persons and assisting environmental experts

Level	Responsible person	Assisting person
Board	Board member with environmental responsibility	Environmental manager at corporate level
Group/plant	Group/plant manager	Environmental manager of group or plant
Process	Process manager	Environmental process expert
Product	Product manager[a] (often the group manager)	Environmental product expert (might be the same environmental manager as at group level)

[a]The real responsibility for the product is usually on the highest – the group level, or with SME on the board level! Product managers often bear limited responsibility. Plant managers usually have the responsibility for industrial environmental protection

2.5.3   Risk Management

Achieving the guarantee of threshold limits of certain hazardous substances is not easy. It is combined with defined analytical procedures and a random check of some samples. This is similar to checking other physical properties and the energy consumption of a product. As it is part of the incoming inspection, it will be the task of the quality department to conduct the checks. Quality management should check the properties of the materials, but also make sure that corresponding requirements are placed in the software for orders. The difference between the identification of normal failures and environmental ones is that, in the case of an environmental "failure", the company may be in violation of the law. So an early warning system is required that combines experts from quality, environment, and technology. The information from the supplier must be gathered and evaluated. Required properties must be systematically integrated in contracts.

A new task for a quality engineer can be to evaluate risks for the company from environmental failures in purchased goods, components or materials, overly delayed development, or new legislation. A management system concerning environmental risks will include reporting, potential costs, and an action plan. Forthcoming legislation should be investigated, as should international regulatory action, standards under development, and environmental progress of competitors. An example for the risk assessment procedure can be found in Annex 4. In Annex 5, an example for the cost risk calculation combined with forthcoming legislation is given.

The process to make sure that no prohibited substances will be used in a product should require: (i) that suppliers answer questionnaires, (ii) inclusion of the requirements in contracts, (iii) material testing, and (iv) sample preparation if necessary. Also, review steps should be integrated into every product development scheme. Some standards for that purpose are under development or have already been developed by IEC TC 111 [e.g. sample preparation; hazardous substance management; analytical procedures]. Because of complexity, data management must also be installed as part of the management system. A checklist for the requirements of a material management is added in Annex 6.

Checklist for Management Analysis

Assessment questions	Answer	Comments	Recommended follow-up activities	Related sections
Do you have environmental management systems in place?	Yes ☺ No ☹	ISO 14001, EMAS	International extension	2.4, 3.5
Is Ecodesign already included in a management system?	Yes ☺ No ☹	IEC 62 430, Ecodesign directive Annex IV/V, in ISO 9001 or ISO 14001, integrated management system	Review	2.3, 3.5

(continued)

Checklist for management analysis (continued)

Assessment questions	Answer	Comments	Recommended follow-up activities	Related sections
In management systems, are missing environmental processes identified and installation planned?	Yes ☺ No ☹	Energy, recycling, communication, chemicals, software	Continuous extension	2.3, 2.4, 3.5
Are missing environmental tools identified and application planned?	Yes ☺ No ☹	LCA, EPD, Design rules, PCF, EQFD,	Extension	2.3, 3.3,
Are environmental responsibilities checked at all levels and are responsible persons nominated?	Yes ☺ No ☹	Responsibility at management level, expert-level assistance required	Assign responsibilities also for new processes	2.3, 2.4, 3.5
Is a document available describing all environmental tasks, responsibilities, and processes?	Yes ☺ No ☹	Environmental principles are also a guide for customers and the public	Annual check of validity	2.3, 2.4, 3.5
Is quality management involved in checking environmental properties of components, materials and products?	Yes ☺ No ☹	Incoming tests required, specifications required, Integration in software systems, contracts required	Update regularly	1.2, 2.3, 2.4, 3.5
Have sources been identified for systematic information about trends in legislation and standards?	Yes ☺ No ☹	Sources could be associations, consultants	Continuous participation	1.2
Has risk management been installed for environmental risks?	Yes ☺ No ☹	Includes trend analysis with legislation, standards, competition	Annual update	1.2, 2.3, 2.4, 3.5
Is an early warning system in place to identify environmental problems with components or materials in time and to inform the management?	Yes ☺ No ☹	Problems can come from suppliers, from public, and from media. Authorized persons must be available to manage a crisis	Test system	1.2, 3.5

2.6   Conclusion of Situation Analysis

Results of the afore-mentioned sections might give answers to the following questions: Which product fits the market best? What does the market expect? Which is the environmental part of the product profile? What should be expected from suppliers as additional support? What is missing in the management system?

But before functionalities are selected, a check for plausibility should be undertaken:

– Where are the contradictions?
– What does not fit together (functionalities, properties)?

Explanation

Easy disassembly of valuable components for reuse must be managed if their reuse is being planned. Planning for several product generations will frequently facilitate higher rates of reuse. All the requirements have to be counter-checked one by one.

– Which functionalities are required for the product and what is more or less combined with the application in the corresponding system?

Explanation

A refrigerator does not only supply cooling but also storage with slow fermentation. In Korea, a refrigerator provides storage and slow fermentation for "Kimchi", which is a fermented vegetable similar to German Sauerkraut. In Korea and Japan, washing laundry in cold water is included in the washing cycle, compared with washing in Europe where water temperature can be as high as up to 90°C.

– What does the system approach really mean for my product? Are there several systems into which my product fits?

Explanation

A System approach means understanding the role of the application. For the example of the *instabus EIB*® (a controller product family for all energy-consuming applications in a house) it means that with some modifications, the product can be applied to many functions like lighting, alarms, kitchen controls, shades, switching or dimming in the "house" system. Very similar products can be designed and changed in function by software. By using a system approach, products better fit the requirements of the customer and lessen their environmental impact like non-hazardous flame retardants, lower incineration load, less wires from the main panel. If product families are developed, the environmental impact can, in total, be minimized as compared with the isolated development of individual products, each with their own special function. For an old building, installation costs can be high. The product, as a controller, could however, fit other systems such as those in industrial buildings.

– Which scenarios would be valid for the future development of the system and which consequences could occur for my product?

Explanation

With scenario analysis, the future opportunities of a product can be estimated. Which modifications or services might be of interest? In a special case, a question might be: How many homeowners could install this relatively expensive control equipment of an *instabus EIB®* ? How much energy reduction is possible if this kind of equipment is applied worldwide? What promotion is possible?

– Which product fits the market best? What does the market expect?
  After comparing an existing or planned new product with those of the competitors, potential new functionalities can be derived for the final new product.

But before taking the final decision, further questions should be answered:

(a) In which growing markets and with which service can the product be placed?
(b) Does the product fit with general trends, like energy reduction, less hazardous substances or less waste?

Explanation

An interesting example is the trend toward "megacities". This trend creates a potential new target group of customers who will demand better energy efficiency, more networks, cooperation between products in a system, a focus on installation, mobility, a need for environmental protection, and, as always, a low enough price. Would the new product fit into such a trend? If not, perhaps the company should change its mission, for example, from a simple switch producer to a company selling everything connecting people in a megacity. Such a new business model could be much more compatible to the environment and could offer new business opportunities.

(c) After the evaluation of alternatives, a rough estimate of a favorable solution might be possible. From that, the environmental part of the product profile should be derived and given as an input to the overall profile.

Some questions to consider:

- Which is the environmental part of the product profile?
- Describe the necessary attributes and functionalities of a product.
- Where are the competitors?
- Which values should be reduced and how much?
- Which attributes belong to which part of the life cycle?

Answers to these will provide a list of required attributes and services. A cross-check should find contradictions with other requirements such as technical features or target costs.

(d) What should be delivered by suppliers as additional support? Proof of an environmental management system and absence of environmental and other risks is needed. Suppliers should especially develop components with very low energy consumption. Similar kinds of information on compatible software should be provided to suppliers, such as a list of materials or required properties. Cooperation on reuse, including testing of components, could be contracted.

(e) Is your management system up-to-date? Many new tasks need to be managed, from Ecodesign to the management of risks. Start with a list of new tasks and corresponding processes. All missing pieces should be described and evaluated to determine what is really required and how the missing pieces are to be integrated into the company's systems.

Chapter 3
Strategy Development

After the detailed analyses undertaken in Chapter 2, it is time to decide on the kind of strategy that will help your company achieve competitive advantage through Ecodesign. In doing so, one should examine the corporate, market, product, production and management aspects of your business and design an integrated strategy.

It should be emphasized that, while developing a good strategy always requires a thorough understanding of a given situation, it also requires enough flexibility to manoeuvre in those areas where potential action might be better, if taken later.

3.1 Corporate Considerations

Positioning the company in the marketplace is one of the most important strategic issues of corporate level management. The desired view of the future is, in essence, where the company would like to be in regard to its environmental performance 5, 10, or 20 years hence. The basis for this work was laid with the trend analysis in Section 1.2 and the corporate analysis in Section 2.1.

As a first step, the environmental mission and vision of the company should be reviewed to ensure that it integrates sustainability, environmental aspects of the whole life cycle, co-operation with customers and suppliers, and perhaps also international behaviour. An example of one company's environmental policy follows:

W. Wimmer et al., *ECODESIGN – The Competitive Advantage*,
Alliance for Global Sustainability Bookseries 18,
DOI 10.1007/978-90-481-9127-7_3, © Springer Science+Business Media B.V. 2010

Example: Policy for Environmental Protection, Health Management and Safety 2010 (Siemens)

As a global company we are facing special responsibility for worldwide long-term challenges such as demographic change, climate change and diminishing resources. Sustainability is the key to securing our company's future. Our commitment to being a socially responsible company requires that we meet the demands of commerce in an ecologically and socially sound manner. Hence, achieving excellence for Environmental Protection, Health Management and Safety is a high priority within Siemens. A demonstrated commitment will increase the competitive advantage of businesses and our customers, and lay the foundation for a successful future.

Siemens strives through out innovative products, systems and solutions to improve the quality of life the world over. This includes high energy efficiency for climate protection, supply of clean water, health protection, and environmentally compatible transportation systems. For this, we consider the entire product lifecycle.

We design, develop, manufacture and market our products and services so as to protect the environment and human health in a manner that meets or exceeds any applicable regulations, and in order to minimize the impact on our natural resources. We design our working conditions in ways that safeguard our employees' performance, safety, health, motivation and satisfaction.

Environmental Protection, Health Management and Safety contribute towards human health and the company's assets. Our global system for managing EHS is the basis for continuous improvement of our performance on these areas.

All managers and employees act in accordance with this policy and observe the relevant regulations [1].

A review may or may not result in any changes to your company's mission and vision statements. However, *strategic environmental targets* should be developed for *production* that take into account (a) sustainability of processes; (b) emissions; (c) energy; (d) waste and water reduction; and (e) quantity of renewable energies.

Similarly, *strategic product targets* should be developed that take into account: (i) leadership with environmentally compatible products; (ii) avoidance of special hazardous substances; (iii) reduction of energy consumption; and (iv) contribution to governmental energy reduction targets.

Alert

For setting future targets, it is not necessary to conduct difficult scenario analyses. Many companies have displayed their outlook for the next 20 years on their homepages. Also, governments such as the EU, have published long-term targets or studies from which targets and necessary changes can be derived.

Don't be concerned about proposing production related energy reduction targets such as 20% to 30% within 5 years. It is possible! Values are relative, and relate to certain levels of product turnover. Similar improvements can also be achieved for products [1].

An environmental rating of the products might also be of interest. In many cases, cost reduction targets can be combined with environmental ones. The Boston Consulting Group proposed in the 1990s that the environmental properties of products will become the third decisive factor in competition together with cost and quality (cf. Introduction).

Depending on the targets set, your company can become a "green" one or a "me too" company. In the second case, environmental properties are only fulfilled where legally required. If the company wants to become a market leader, the environmental positioning will need to be specified. For example, a company can choose to become a "zero carbon company" or a "zero waste company" where it is shown that the total contribution of the company to global warming or waste is zero [2] or the company can set several targets [3]. However, such a strategy is not always without risk, and accusations of "green washing" might arise, especially if targets are vague or not actually achievable. It might be better to show overall contribution and reduction potential. Japanese companies, like Sony, have changed their strategy from "zero carbon" 10 years ago, to now using more realistic and achievable targets [3].

The international policy of any company should refer to the same standards worldwide, including environmental management systems in every plant, take back of waste products, and reference to international reduction targets. A communication strategy should reflect the success already achieved. More and more companies like GE or Siemens communicate now the percentage of their portfolio which consists of "green" products.

Often, this kind of aggressive improvement strategy creates a culture of innovation within the company, involving employees, and often including a proposal system for innovation. Suppliers should also be involved early in any new developments, including those suppliers that supply components that have the potential to save energy. It is worth noting that environmentally improved components are more likely to be available when they are requested by the customer rather than when the customer relies on chance!

Checklist for Corporate Considerations – Section 3.1

Check for Strategic Considerations

Assessment questions	Answer	Comments	Recommended follow-up activities	Related sections
Have you decided which corporate level changes are required to achieve necessary future environmental performance?	Yes ☺ No ☹	From scenario planning, consumption levels of products (e.g. in 10 years) can be estimated, which can lead to a 10 year action plan	Adjust scenario (e.g. with customers and suppliers)	1.2, 2.1, 2.2, 2.5, 4.1; 4.2
Have you decided what is to be included in a new mission and vision?	Yes ☺ No ☹	Environmental positioning means shifting the company to sustainability	Check whether employees accept and live it	2.1
Do you have targets for reducing production related environmental impacts?	Yes ☺ No ☹	Targets are required to facilitate better environmental performance	Annual reporting	3.4, 3.5, 4.4, 4.5
Are targets set to reduce the impact of all products sold in the market?	Yes ☺ No ☹	Total impact is politically important. Products with very small sales are less important	Annual reporting	2.1, 2.3, 4.3
Have you determined your green "flagship" products with which you can advertise?	Yes ☺ No ☹	You can also include reduced impact from use of renewable energy	Show annual progress	2.3, 3.3
Have you decided in which markets "green" products should be promoted?	Yes ☺ No ☹	General marketing strategy for environmentally compatible products is required	Implement program	3.2, 4.2
Has the company defined which environmental targets contribute to (a) cost reduction, (b) risk reduction, and (c) improved security of resource supply?	Yes ☺ No ☹	Impact reduction often also means cost savings. Risks occur if there is No reaction (e.g. to legal trends). Scarcity of materials can cause high costs	Build up environmental reporting with costs, risks, improved safety	3.5, 4.5
Are statements on what your company has already achieved ready for communication?	Yes ☺ No ☹	Public awareness of your progress is important	Extend statements (e.g. to free of … lead, asbestos, etc.)	4.1, 4.2

Of course, more is always possible! Ambitious targets cannot be achieved through continuous and incremental improvement alone. New developments will need to be launched, including the discovery and development of entirely new technologies.

To get a sense of environmental trends, contact with different environmental associations might be worthwhile. Also, by participating in an international rating system, such as the Dow Jones Sustainability Group Index (DJSGI) [4], some insight into future requirements might be obtained. Both can provide feedback on how your company is viewed by those outside your organization. How your company will be viewed in the future depends on long-term programs and commitments and the trust your customers have in your products.

3.2 Market Considerations

In Section 2.2, a series of questions were posed about the market sector in which your company and its products are situated. The purpose of this section is to help you decide, first of all, whether to engage in any environmental marketing at all and, if so, to determine the right kind of strategy.

For the purposes of developing an appropriate environmental marketing strategy, the following are important issues. They relate to:

- The size and level of your direct competition
- The degree of segmentation of your market sector
- The environmental trends acting on and influencing your market sector
- The degree to which your competitors are active in environmental marketing
- Whether your competitors' environmental messaging is well received
- The level of customer interest in your product's environmental attributes

3.2.1 Pros and Cons of Environmental Marketing

For some companies, the use of environmental factors in their marketing strategy is deliberately avoided. The reasons for this may vary, but the following appear to be the most common:

- For companies that manufacture or distribute different products in the same category that compete with each other (usually consumer goods products), they may not want to have some products identified as environmentally superior, thereby leaving the impression that they are also selling environmentally inferior products.
- Some companies are concerned about promoting the environmental aspects of one of their products, thereby creating the impression that the whole company is an environmental leader.

- Some companies may be uncomfortable about promoting the environmental virtues of their products, while their manufacturing facilities (either domestically or abroad) may be operating at a low level from the perspective of sustainable development. This could include human rights issues or pollution issues in less regulated countries.
- Some companies have experienced high levels of scrutiny after proudly advertising the environmental progress of a specific product group. Advocacy groups then demanded more information about other aspects of the products in question, as well as the company's other products and overall corporate operations.

In summary, the primary reasons that a company may choose to avoid environmental marketing is concern about raising expectations and avoiding time consuming, often undue, and certainly unwanted scrutiny. However, choosing the avoidance option may be a short-sighted view. As noted in previous chapters, customers are increasingly interested in the environmental performance of the products they buy and this trend will impact on your sector sooner or later. Furthermore, governments in many regions and countries are now also looking at the environmental aspects of products. Many governmental agencies, if they have not done so already, are considering how they can accelerate the reduction of environmental impacts arising from product manufacture and use.

All of this should lead a company to the conclusion that (i) product related environmental performance needs to be improved and (ii) promoting this progress will give comfort to both consumers and governmental agencies that action is being taken. As well, this type of initiative could lead to competitive market advantage.

3.2.2 Elements of an Environmental Marketing Strategy

Effective marketing begins with an examination of the problem or opportunity, continues with a review of how the opportunity might be exploited, examines the motivations and needs of the client, and then designs and positions the product, concept, or service accordingly. The main elements of any type of marketing strategy are

- Assessing the target audience for the marketing efforts
- Getting the message right
- Ensuring that the right media are employed

3.2.2.1 Target Audience

In this section we apply these basic marketing practices to environmental challenges through an examination of the motivations and needs of the client or audience.

Marketing has always been concerned with consumers' motivations – that is, what drives people and organizations to demand and, ultimately, purchase different goods and services? What do they want? If a purveyor of goods or a service knows what is being called for in the marketplace, then there is a much greater likelihood that the good or service will be purchased and that his business will thrive.

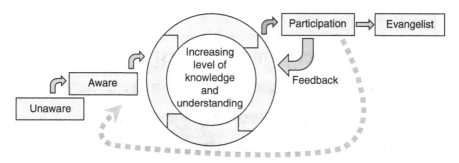

Fig. 3.1 Sustainable motivation

The following schematic (Fig. 3.1), taken from a UNEP report on "Sustainable Motivation" provides an interesting view of how consumers' values can progress from "unaware" to "evangelist".

The authors argue that, "from the perspective of motivation, we need to understand why people make the move illustrated by the pink arrow – that is, from knowledge and understanding to participation. Crucially, it is because they have gone round the loop of increasing knowledge and understanding sufficient times and that an opportunity arises for them to take that step. Much of what we need to know about motivation will be rooted in how that knowledge and understanding is created. People will be at different stages of this cycle…"

This is similar to the view that real education is the three-step process of: (i) raising awareness; (ii) enhancing understanding; and then (iii) seeing action.

Marketing has a potentially key role to play in helping to educate consumers. However, traditionally, much "environmental marketing" has concerned itself only with efforts to try to identify so-called "green consumers" – that is, that segment of the overall population that is most likely to buy those products with some "green" or "environmental" advantage. Identification of these green consumers originally focused upon isolating unique demographic characteristics – speculating that wealthier, female, more educated, and younger individuals were those who should be targeted.

In the 1990s, however, strategies to isolate these green consumers began to prioritize attitudinal characteristics that revealed those with particular "worldviews" (like liberalism) and socialization patterns that uncovered people who were connected to relevant social networks. While certainly more sophisticated, these efforts continued to prove relatively disappointing, with green marketing efforts showing little improvement from the original modest level of accomplishments. Indeed, we would argue that marketing efforts on environmental issues have generally fallen into the trap of targeting those who are terrified of dramatic environmental damage; want to 'do the right thing'; or want to save money. This is an incomplete approach. Figure 3.2, from a TerraChoice Environmental Marketing Inc. presentation, provides some detail on the spectrum of environmental interest.

This demonstrates that about 20% of consumers have a strong disposition toward the environment and would be most likely to take action on that front.

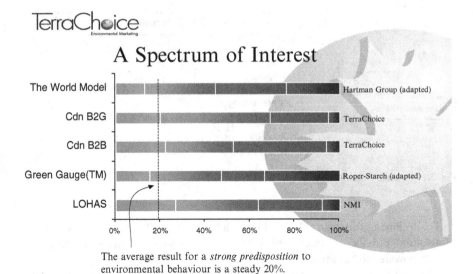

The average result for a *strong predisposition* to
environmental behaviour is a steady 20%.

Fig. 3.2 Shades of Green – a spectrum of Interest

We can likely conclude that the "green 20%" represents the kinds of consumers
that are traditionally being "targeted" with the "we are going to die" and the "right
thing to do" kinds of messages. If this 20% is correct, then some 80% are left out
if marketing efforts were focused only on the green end of the spectrum. However,
economic self-interest and "self gratification" are most probably in the interests of
both the "green" 20% and the remaining 80%.

We therefore think that it is time to try something new. That "something new" is
to move the focus away from the "green consumer", and to consider "all consumers"
as the potential market for environmental products, activities, services, and facilities.
To do this, we review the ideas of Abraham Maslow in order to reflect upon peoples'
and organizations' motivations in a general sense. We acknowledge, of course, that
there is nothing particularly original in such a review, for Maslow has been a mainstay
of marketing studies for many years. Where there has been much less attention,
however, is the investigation of the way in which Maslow's ideas about the "hierarchy
of needs" might be applied directly to strategies for environmental marketing. This
is the new area that we hope to open up for further discussion. But, first, we present
a brief review of Maslow's hierarchy of needs.

In 1943, psychologist Abraham Maslow wrote a paper entitled "A Theory
of Human Motivation", in which he proposed that there exists a hierarchy
of human needs. He postulated that as humans fulfil basic needs, they then
desire higher needs. Maslow's original work – and the ways in which he
subsequently elaborated it – has formed the basis of much consideration of
what catalyses, directs, and/or generates human behaviour.

Maslow argued that the hierarchy of needs consists of five levels, often (as we present in Fig. 3.3) depicted as a pyramid. The four lower levels were grouped together as "deficiency needs", while the top level was called "being needs". Maslow also argued that the higher needs come into focus only once the lower needs have been met. Before, for example, the need for love/belonging can be part of one's desires, one's physiological and safety needs must have been fulfilled. Let us briefly review what Maslow meant by each of these five levels.

Once physiological needs are met, people turn to their "safety needs". These are the needs for security – the need to ensure that the individual is not in danger from external forces (for example, wild animals, criminals, natural disasters, etc.).

The third layer of human needs consists of the desire for relationships with others – perhaps friends, lovers and/or a community in general. These are referred to as the "love/belonging" needs, and they encompass the requirement for humans to belong, to be accepted by others.

The highest of the deficiency needs is the esteem needs. The dictionary definition of "esteem" is 'favourable opinion, regard, respect' [5] and Maslow suggested that this level of needs included both self-esteem and the esteem of others. It, therefore, is in addition to simply 'belonging' to a group (part of the third level) and is suggestive of having the respect of others.

Finally, actualization – or, "self-actualization", as it has often come to be called – refers to the intrinsic needs of humans to make the most of their unique abilities. It is, at its simplest, the need to "be what you can be". Maslow, himself, added to this level in subsequent work, introducing the term "self-transcendence", to suggest that this is all about connecting to something beyond the self (beyond one's own ego) or about helping others fulfil their potential.

Maslow's ideas, though widely used, are not without their critics. Although we do not review the broader debate about the value of Maslow's work here, it is not our intention to suggest that the "hierarchy of needs" should necessarily be accepted

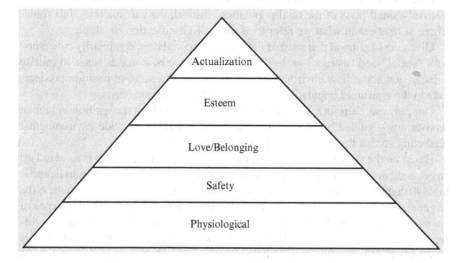

Fig. 3.3 Maslow's hierarchy of needs

"as fact". Instead, it is our desire to use the levels introduced by Maslow to generate additional ideas about how strategy surrounding environmental marketing could be developed.

Above, we make the point that most environmental marketing has targeted relatively small portions of the overall population – that is, it has searched for the "elusive green consumer". Indeed, many of the formalized scales used to classify the population into different groups have been used by marketers of green products. Frequent Green Gauge Reports by Roper Starch Worldwide [6] for example, try to identify so-called "true-blue greens", with the message being that targeting this segment of the population will yield success in selling environmental products.

Indeed, we can even interpret Maslow's work in a similar way. More specifically, much environmental marketing to date has seemed – consciously or unconsciously – to focus upon either end of Maslow's hierarchy of needs. At the bottom, environmental product messages that aim to strike fear into the hearts of consumers effectively serve to suggest that their most basic physiological and safety needs are not being met. At the top, meanwhile, those messages that encourage people to act for altruistic reasons are, in effect, offers to meet consumers' actualization needs. The example of the "what would Jesus drive" campaign is an example of this.

In reality, however, most people in industrialized nations are operating at neither the bottom nor the top of Maslow's spectrum. Most generally, they have enough to eat, and they do not live in constant fear of being attacked on the street. With these first two sets of needs having been met, therefore, most people can consider higher levels of needs.

Nevertheless, most people have fulfilled neither their love/belonging needs nor their esteem needs, thus not allowing them to focus their energies upon meeting their actualization needs. Instead, the majority of the population is operating in this middle ground: people are interested in figuring out how they can develop supportive family relationships, cultivate quality friendships, gain the respect of peers, and so on. Indeed, we speculate – as per the notional population distribution in Fig. 3.4 – that those trying to fulfil "survival needs" and "self-transcendence needs" make up relatively small parts of the total population. Instead, the vast majority, fall somewhere in between in what we refer to as the Self-Gratification envelope.

Given the likelihood that most of the consumer population is primarily concerned with "social" and "esteem" or "self-gratification" needs, it makes sense to initiate a discussion about how environmental marketing could serve to position products to help the consumer population meet their self-gratification needs.

Our purpose, therefore, is to stimulate thinking about how recognition of human motivations – guided by Maslow's hierarchy of needs – can guide environmental marketing so that the entire market potential is considered.

Based on the above discussion, one can easily conclude that there is no standard target audience. If a company is in the consumer goods field, even this audience is highly fragmented. For example, if the product category at issue is cosmetics, the target audience is most likely women and, within that audience, different types of cosmetic products might be aimed at either different age or ethnic groups, or different occupation categories. However, if the product category is yard care or "do-it-yourself", then the target audience is most likely men who live in houses and are between the ages of 25 and 75.

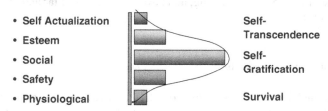

Fig. 3.4 Notional population distribution

However, the purpose of identifying the target audience is to determine its main likes, dislikes, orientation, and any possible predisposition. This in turn helps to better craft the message. In terms of environmental marketing, the kinds of things that should be reviewed in relation to target audience include the likely age grouping for the product category, this group's environmental orientation in terms of knowledge, interest, and possibility of being swayed by environmental messaging.

3.2.2.2 The Environmental Message

Once the target audience is adequately identified, the next step in the strategy is to develop the right message. Here, consideration needs to be given to whether the messaging aims to:

– Inform the audience about the environmental aspects of the product or
– Identify that the product has environmental leadership characteristics relative to competing products

Other considerations are whether or not (a) an independent party plays a role in confirming the voracity of the message, (b) all aspects of the life cycle are included; or (c) what kind of label, if any, should be used.

Regardless of the decision, the following principles, which are taken from the study by TerraChoice Environmental Marketing Inc. on the "Six Sins of Greenwashing" [7] are useful guides:

(i) Be cautious about highlighting one environmental attribute while ignoring other potentially more significant environmental attributes.
(ii) Ensure that any claims are, as a minimum, verifiable. Often, the best approach is to have an independent party verify claims.
(iii) Avoid any claim that is non-specific or vague. A term such as "natural" or "chemical-free" can be both true and false depending on interpretation and context.
(iv) Ensure that claims are relevant. Claiming that your home ink-jet printer is free of DDT is technically correct, but there are no printers made with DDT.
(v) Similar to the first principle, avoid promoting the environmental aspects of a product that is, by definition, harmful to the environment, just a bit less so than

competing products. A company's cigarettes will not be considered green just because they use less packaging, and have lower levels of tar and nicotine.

(vi) Be truthful.

3.2.2.3 Choosing the Medium

A range of media could be used to promote the environmental aspects of your products. These include the usual print, radio and television advertisements, in-store messaging tools (shelf-talkers, etc), and product labels. Between the type of target audience, the product category, and the kind of messaging approach chosen, the appropriate media will probably become more obvious. For example, if

(a) The product category is home office equipment (e.g. printers, computers, fax machines)
(b) The target audience (i.e. the primary purchasers) are home owners who are interested in energy conservation and indoor air quality
(c) The message is that your products are energy leaders, have very low indoor air quality impact, are easy to disassemble, with many reusable components
(d) You have chosen to use a third-party environmental leadership certification

then, the appropriate media could begin with a certification label, and might include press releases about achieving the certification, and in-store promotional materials.

3.2.3 Making the Right Choice

Table 3.1 [8] highlights several marketing considerations for three types of environmental labels, Types I, II, and III as defined by ISO in their 14020 series of standards. Of course, there are many more and different kinds of labels in the market today, so these are provided only by way of example.

In summary, there is no one right environmental marketing strategy that can be universally employed. While there is a process that one can follow to help get to the right approach, there may be risks. This section has hopefully helped you develop an appropriate environmental marketing strategy in full knowledge of both the opportunities and the risks, and thus helping you mitigate any risks that have not yet been addressed.

3.3 Product Design Considerations

... an existing product should be redesigned, but this time environmental issues should be taken into account ...

... the market is demanding products that are designed to better address environmental issues ...

Table 3.1 ISO label types

Different types of environmental labels and declarations, advantages, disadvantages and application areas [Lee and Uehara, 2003]

Item	Type I	Type II	Type III
Generic Name	Eco Labeling	Self-declared environmental Claim	Environmental product declaration
Target Audience	Retail Consumers	Retail/ industrial/ institutional consumers	Industrial/ Institutional/ Retail Consumers
Communication Method	Environmental Label	Text and Symbol	Environmental Profile Data sheet
Scope	Whole life cycle	Single aspect	Whole life cycle
Use of LCA	No	No	Yes
Advantage	Easily identified	Market oriented	Detailed data via common method
	Quick decision	Flexible approach to market needs	Credibility via scientific quantitative data
	Credibility through third party	Tool for inter-business competition	
Disadvantage	Only uses a symbol (logo)	Relatively low credibility	Complicated LCA analysis
	No detailed information	Need to face consumers directly (no third party)	Insufficient back ground data
	No linkage to company's unique effort	Claim is about a single issue or limited	Not easy to comprehend
Application area	Home use products/ simple function products/low priced products	Products in general	Products for industrial use/relatively complicated and high priced products/durable products

Checklist for Market Considerations – Section 3.2

Assessment questions	Answer	Comments	Recommended follow-up activities	Related sections
Is your target audience aware of environmental issues?	Yes ☺ No ☹	Feedback from customers may provide part of the answer	Research or conduct market analysis	2.2, 3.2, 4.2
Is your competitor already using environmental marketing?	Yes ☺ No ☹	If yes, check to see whether there is any traction (i.e. success) as a result of the efforts?	Determine what the customers for your product category want	2.2, 3.2
Is the environmental message you want to communicate clear?	Yes ☺ No ☹	Also, check the message against the principles in 3.2	Conduct focus groups to determine clarity of message	4.2
Are the environmental marketing instruments chosen already?	Yes ☺ No ☹	If yes, check to ensure that results will deliver what you expect	Focus groups can help choose the optimal instrument	4.2

... a competitor is working to have the first ecoproduct on the market ...
... the B2B market is requiring good environmental performance in tenders ...

You may have experienced one of the above situations or found yourself in similar discussions during a product development process. So, how to start? Where and when should you start including environmental issues into the product development process? Ideally, the best time to consider environmental improvements is when developing a new product, or redesigning an existing one. The earlier environmental issues can be considered in the product (re)development process the better.

Integrating environmental issues in an appropriate way should begin with a list of product specifications, which is the usual starting point for product development (see Fig. 3.5). The key point is that environmental considerations are addressed during this very early stage in the process when there is still enough "room to manoeuvre" should new requirements arise, such as those related to environmental perspectives.

The question now is how to translate "environment" into more familiar terms for product developers, and which can be taken into account in the list of product specifications. What is important for the product being designed? Is it a high recycling rate, a long lifetime, optimized packaging, low energy consumption during use, light weight design, avoidance of certain problematic substances, or ... ?

Well, you may say that probably all of the ideas listed are valuable and should be considered. That's probably correct, but, realistically, there is often not enough time or resources available in the product development process to address all of the possible environmental performance improvements. As such, the most significant ones will need to be selected. Effective methods to guide this selection are presented in Section 2.3.

The whole process is similar to cost improvements. There are, most likely, many possibilities leading to lower costs, but those that result in the most significant reductions should be considered for further follow-up, at least initially.

Fig. 3.5 Product development process [23]

To assist in knowing what is "most relevant" and how to follow up on those, three steps are proposed:

- A *pre-study* on the environmental performance of a previous or similar products.
- *Staff training* to help everyone better understand environmental issues needs to be done up front. A successful improvement process is based on awareness and understanding of the environmental situation and context.
- *Support* to work out improved environmental performance of products is needed. Appropriate tools and methods for easy integration in the product development process are required as well as a good environmental data source for decision-making during product design.

3.3.1 Pre-study

The purpose of the pre-study is to gain insight into the most relevant product-related environmental issues. Using the methods described in Section 2.3, the environmental profile (i.e. the distribution of the environmental impacts over the product life cycle) needs to be worked out. In doing this, we typically face two challenges: (i) life cycle data from the product we are going to design may be very limited; and (ii) most likely, the time to perform the pre-study is limited.

As a consequence, the best way forward is to choose a similar product, either a predecessor model of the product, or another product that has a similar product life cycle (use of raw materials, manufacture, distribution, use, and end of life). Once this "representative" product is found, the "right" method has to be chosen. Since, at this early stage in product development, the aim is to define the right improvement strategy, there is no need for a time consuming LCA according to ISO 14040, or any other detailed assessment of environmental performance. The level of detail in doing the environmental assessment should be appropriate to the intended result. Working with assumptions is fine at this point. A quick evaluation is sufficient where we aim at being "approximately right instead of exactly wrong". "Approximately right" refers to using assumptions in case there are not enough data, such as assumptions on the way the product will be manufactured (considering main production processes) or the way the product will be distributed (referring back to the distribution of a similar product).

The use stage of a product needs special attention when it comes down to defining the use scenario. How and in which way will the product be used? How frequently? What is the energy or material consumption during the use stage? If there are uncertainties, and there will be, one can propose and test different use scenarios like "intensive use" or "less intensive use" to see the differences in outcomes, and to evaluate the consequences for the environmental impacts in the use stage. A practical example would be the use of a scenario for a washing machine for a family household compared with a household of a single person. One application could easily have up to four to six times more "intensive usage", resulting in a much larger environmental impact. Consequently, the environmental design strategies may be different for these two different types of washing machines.

The pre-study process results in the environmental profile of the product and from that the key environmental performance indicators (KEPI) can be derived. At this stage, it should be clear what we are aiming at in the product development process. The vague starting term of "environment" is now translating into, for example, "reduction of energy consumption during use", "product weight reduction", or any other strategy depending on the KEPI.

3.3.2 Staff Training

The "pre-study" could be conducted by an environmental leader within the organization. However, if it is the first time such a study has been done, it is advisable to seek external support, especially when there is limited knowledge or time within the organization to run the pre-study. To be successful with the redesign of the product, the product development staff should receive training and motivation in order to improve the environmental performance of their own products. This is a critical issue, since improving a product always means identifying and dealing with weaknesses. This knowledge transfer has to be implemented in a sensible way, avoiding resistance due to criticism of previous work done, often, by the same product development team.

It is recommended that workshops, or other awareness and training activities with foreign products not developed by the trainees, be conducted. This way, it is much easier to get acceptance and to identify weak points.

The overall aim is to develop insight into the overall environmental context, identifying why certain issues are more important than others, and also linking the potential activities in the development process (in terms of milestones and deliverables). Here, staff members may need some guidance. Ideally, at the end of such a training process, the basic connections between the environment and the product's life cycle are established, potential activities are identified and linked to the product development process and simple methods and tools are introduced to support daily work.

3.3.3 Support

The whole process of calculating the environmental profile of the reference product requires practical tools and reliable environmental data. Support in decision making is also needed later in the development process, when searching for and implementing new solutions. What are the environmental consequences of new or alternative design decisions? Getting to understand this, but also being able to plan for enhanced environmental performance, is the key to supporting design teams. Certainly it is not all about environment. Thinking in environmental terms is an additional task for the design team and, therefore, support is required.

Simple "if this – then that" rules, easy to understand evaluations, or key values and data are helpful at this stage. Complicated assessments are often difficult to accept and may even be inappropriate. Environmental data should be prepared beforehand and communicated during staff training. Also, some practical exercises should be conducted to get staff comfortable with working with these data and tailor-made indicators. There are no general rules on how to prepare these decision support systems, but it is advisable to take into account the characteristic of a product and its environmental profile. Useful indicators for a raw material-intensive product could be CO_2-equivalent per kg of material or energy values in MJ per kg of material to see the difference a decision for an alternative material could make. Similar data should be prepared for other processes in the product life cycle.

Using that set of environmental data, the design team can not only identify or calculate potential improvements but also check whether the improvement in one stage in the life cycle is going to affect the impact at another stage. In the product redesign, shifting the environmental impact from one stage in the product life cycle to another stage must be avoided.

3.3.4 Product Characteristics

The pre-study identifies the basic characteristics of a product. Two possibilities exist: one, where the use stage in the product life cycle is most relevant; and the other, where the use stage of the product is not relevant, but the use of raw materials and/or the manufacturing stage are important. Certainly, there are exceptions, but most of any product's environmental impacts can be found in these two scenarios (see Figs. 3.6 and 3.8).

3.3.5 Use-Intensive Products

A use-intensive product is a product whose main environmental impact occurs during the operation of the product. This arises either from consuming energy or materials (or both) during use. The environmental impact in the use stage is usually 10–20 times higher than the impact from the use of raw materials or during the manufacturing stage (Fig. 3.6).

Such use-intensive products need to be improved by carefully looking at the user's behaviour and the specific use scenarios. Design for optimal use, and reduction of energy or material consumption will also be required.

The main questions are:

What does the use-scenario look like?
What is needed to optimize product use?

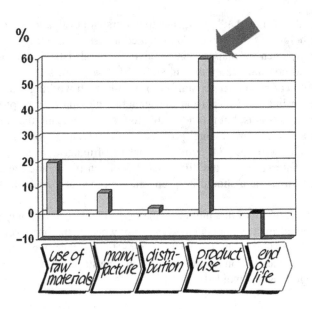

Fig. 3.6 Use-intensive product

Answers to these questions can be found in revising the working principle of the product and in implementing different or new functions to the product. Potential strategies are [9]:

- Optimizing product functionality
- Reducing consumption at use stage
- Avoidance of waste at use stage, etc.

Example: Philips Digital Pocket Memo

A very successful product redesign of a Digital Pocket Memo (DPM) was undertaken at Philips Dictation Systems in Vienna, Austria. The task was to bring environmental knowledge into the product redesign process.

The pre-study showed that the DPM was a use-intensive product due basically to the amount of battery energy required to operate the device. For the intended use over 4 years (4,000 h of operation), about 300 batteries were required. This caused roughly ten times more environmental impact than from the use of raw material and manufacturing stages together.

As a consequence of this pre-study, the improvement strategy looked at the energy system of the DPM. The objectives were (a) to reduce the energy consumption to operate the DPM and (b) to run the device on rechargeable batteries. The resulting redesign yielded a total carbon footprint reduction of 86%. Figure 3.7 shows the predecessor model and the redesigned product.

The improvements were realized through:

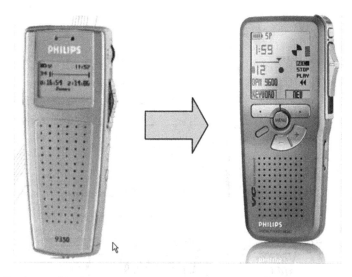

Fig. 3.7 Previous and environmentally redesigned Digital Pocket Memo

- Finding smart energy management allowing 17 h of dictation on one set of batteries (previously 11 h)
- Delivering the device with rechargeable batteries, table stand, and external charger to avoid use of batteries
- Improving product functionality, such as providing USB charging function

Full documentation of the environmental performance of the product has been compiled in the form of an Environmental Product Declaration (EPD) [10].

The resulting new product also provides financial benefits for its user. Since one battery is about one Euro at the current market price, the redesign delivers significant savings.

3.3.6 Raw Material/Manufacturing Intensive Products

Products which have only minor or even no consumption of energy or material in their use stage are most likely raw material or manufacturing-intensive products. An environmental profile of such products is shown in Fig. 3.8.

Potential improvement strategies for "raw material-intensive" products are

- Selecting the right materials
- Reducing material inputs, etc.

But there is a limit to the strategies in reducing the environmental impact due to the kind and amount of materials used in a product. If further improvements are intended, the End of Life stage in the product life cycle will necessarily have to be

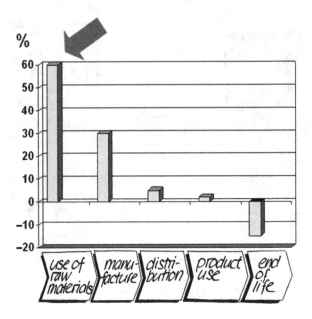

Fig. 3.8 Raw material-/manufacturing-intensive product

considered as well. A combined approach redesigning the Use of Raw materials and End of Life stages will be needed. Of the combined considerations the following product improvement strategies may result:

- Reuse of components
- Recycling of materials, etc.

Subsequently, establishing a resource management system should be discussed and potential competitive advantages identified, especially when looking at the economic benefits of reusing your own components. This may require a different design for most of the products. They will need to be designed such that disassembly can be easily done and processes for upgrading of components can be predicted.

The product structure is also important. An optimized structure is not covered by environmental recommendations alone. However, for recycling, easy disassembly and the chance to get mono-type materials are both very important. In addition, both improvements can provide cost reductions of up to 50%! As mentioned, easy disassembly usually also means easy assembly, and using more environmentally compatible materials (only one type of plastic and one metal) is a well known way to reduce types and parts. But it all depends on the kind of production in which a company is involved. There are three choices:

(i) Identify the product functions required. For a vacuum cleaner, it might be the motor, the housing, the filter, the tube, the brush, etc. These functions are optimised in the form of individual and easily connectable components which also facilitate reuse and repair, if required.

(ii) Combine all parts made of the same or similar materials or which can be made from these materials, where possible, into one or a few parts. This will result in a strong component, material type reduction, and usually cost savings (during storage, logistics, and purchasing). Disassembly is easier when joints are optimised. Materials can be optimized for inscription, recycling, price, and environmental impact. A number of production processes may be eliminated as, for example, only one injection moulding step is necessary for one integrated part, instead of many for many parts.

(iii) If you are in a business like personal computers, it is usual practice to purchase nearly all of the components and the task of your company is to assemble these components. Nevertheless, your chance for improvement is not zero! You can group components to units for simple assembly and disassembly, you can select components that are more energy efficient, or you can optimize the software to deliver some level of energy reduction.

It is certain that nearly all design strategies are hybrid strategies. But considering structural design alternatives also helps to overcome any biases or prejudices that may exist.

All in all, at the product level, we need, to provide the basis of good analysis to determine the best improvement strategies and, at the staff level, with appropriate training and awareness raising to ensure that the new tasks of developing an eco-product are understood. Methods, tools, and checklists are needed to support the product development team in the implementation of Ecodesign.

An ecoproduct can be easily identified when

- Ecodesign is integrated into the early stage of product design and development
- The KEPI and the environmental profile are known
- Design changes result in significant environmental improvements and a shift of environmental impacts from other life cycle stages is avoided
- Environmental improvements are communicated to the market

Furthermore ecoproducts have the potential to attract new business opportunities and deliver competitive advantage.

Checklist for Product Considerations – Section 3.3

Assessment questions	Answer	Comments	Recommended follow-up activities	Related sections
Do you know the environmental performance of the predecessor model?	Yes ☺ No ☹	One should know the Key Environmental Performance Indicators and the Environmental Profile before starting new product development	Perform a pre-study before starting new product development	2.3

(continued)

Checklist for Product Considerations – Section 3.3 (continued)

Assessment questions	Answer	Comments	Recommended follow-up activities	Related sections
Is your staff trained in environmental thinking and decision making?	Yes ☺ No ☹	It is important to develop a good understanding about environmental issues in the product development team	Perform training workshops to ensure a high level of environmental awareness in the development team	
Is enough support provided to ensure qualified decision making in the design process?	Yes ☺ No ☹	Robust data and environmental assessment results, as well as practical databases are needed to make quick environmental decisions in product development	Provide in-house data bases for the most common materials, processes, etc.	
Do you know how to improve the environmental performance of your product?	Yes ☺ No ☹	Determine product design improvements based on the environmental characteristics of your product	Use Ecodesign tools to derive improvement ideas: e.g.: www.ecodesign. at/pilot	

3.4   Production Considerations

3.4.1   Factors to Consider in Production

There are at least four factors to consider in the production of a product. They are

- Resources
- Energy and utility
- Emissions
- Hazardous materials

The environmental goal for the production process is the minimization of the consumption related to the four factors. The resources factor refers to the raw materials of which the product and its component parts are comprised, and to the ancillary materials such as chemicals used in the production processes. The energy and utility factor refers to the energy, water, and air consumed, including electricity for the operation of the production facilities, and for the maintenance of the production conditions such as temperature and humidity. The emissions factor refers to the wastewater, air pollutants, and solid wastes discharged into the environment. The hazardous materials factor refers to the restricted substances and materials used in producing the component parts, either in the product itself or as used in the production processes.

3.4.1.1 Efficient Use of Resources

The consumption of resources is linked directly to the cost and environmental impact of a product. Efficient use of resources reduces not only the cost of the product but also the environmental impact resulting across the entire life cycle of the product. Reducing resources input per product contributes to the conservation of natural resources. In addition, it also helps reduce waste generation.

There are two types of resources: materials for building products and parts thereof (often called raw materials) and chemicals for the various processes, including utilities such as water and air for the operation of the equipment and processes (often called ancillary materials).

Most companies actively seek to reduce the consumption of materials going into the parts and products as part of enhancing productivity. In general, they are less interested in reducing the consumption of chemicals than reducing material consumption. This is because reducing the consumption of chemicals does not translate as readily or as significantly into the level of cost savings achievable from reductions in material consumption. The recent EU regulation on chemicals, REACH, however, will change the attitude towards the management of chemicals in production sites. As such, companies will give greater consideration to the reduced use of chemicals in their premises.

3.4.1.2 Efficient Use of Energy and Utility

Energy (including electricity), liquified natural gas (LNG) and oil, and utilities (such as water, air, cooling water, steam, and vacuum), are all normally used for the operation and maintenance of production facilities. The use of energy not only directly affects costs but also the level of GHG emissions. In particular, the Climate Change Convention and implementation of the Kyoto Protocol and the Copenhagen Accord, including the cap-and-trade system, will cause product manufacturers to set GHG emission reduction targets. This will lead to the reduction of energy consumption. Therefore, companies should monitor the energy consumption activity in their production processes, and develop GHG emission inventory databases by measuring and/or calculating GHG emissions from each and every energy consumption activity. You can use GHG emission databases to monitor and identify improvement opportunities for the mitigation of GHG emissions in your production facilities.

Treatment facilities, such as wastewater treatment plants and air pollution control devices, are an integral part of production facilities. Without successful control of emissions, the entire production process may be at risk of closure. Since the waste emissions depend on the design of the supply and use of the utilities, the design of the utilities must consider the treatment of the waste emissions from the utilities.

3.4.1.3   Minimization of Emissions

Emissions are classified into three categories: Emissions to water, emissions to air, and emissions to land. Emissions are an unavoidable output from any production activity. The real question is how to minimize emissions in the first place, and how to treat the residuals at the lowest cost and highest environmental efficiency.

Two important aspects of the emissions to water are (i) the concentration of pollutants (quality) in the wastewater and (ii) the wastewater flow rates (quantity). Concentration and flow rate are linked to the loss of ancillary materials and consumption of process water, respectively. Therefore, you should implement good housekeeping of the process chemicals and process water in order to minimize emissions to water. In this respect, reusing wastewater can be one viable option.

Of all the possible emissions to air, GHG emissions are currently the most important, at least based on the level of political and public debate. Thus, you should quantify (measure) and manage the GHG emissions from the production sites. This topic is discussed in depth in Section 3.4.2. Another important aspect to consider for the minimization of emissions to air is the separate collection and treatment of the air discharge streams. When a mixture of air pollutants are discharged, the capacity of the circulation pumps and pollution control equipment increases unnecessarily because of the need to handle larger amounts of polluted air.

Emissions to land include solid waste and liquid waste discharges. In order to minimize the emissions of liquid wastes, they need to be collected and treated separately. When solid waste that has a consistently similar composition is regularly discharged, most of this waste can be reused and recycled. In this kind of situation, you should consider the implementation of closed-loop recycling of the solid waste. This not only reduces the emissions to land but also the cost of raw materials.

3.4.1.4   Reduction of the Use of Hazardous Materials

In general, hazardous materials are classified into two groups, prohibited and controlled. Prohibited materials are banned from use in the product and, as such, they are screened out during the purchasing stage. However, prohibited materials and/or substances can sometimes find their way into component parts, although not intentionally added. Therefore, purchased parts and products should be regularly and systematically checked for the presence of the prohibited materials and substances. Controlled substances are not banned for use but may be regulated in the future. In some cases, customers specifically request that their products not contain these materials/substances. Therefore, part and product manufacturers should set up and implement a management or control system that distinguishes the parts containing controlled substances from those containing none. Most ERP manufacturers operate their hazardous substance/material control system under the title of a green supply management system. If your company does not have such a system, a communication channel between suppliers along the supply chain should be established to control the hazardous materials and substances in the parts.

3.4.2   Carbon Footprint

In order to remain competitive in the marketplace, you should reduce the carbon footprint of your products as well as in your company and operations. The cap-and-trade system for carbon emissions will force manufacturing companies to adopt measures to mitigate carbon emissions from their products, production processes, and production facilities. We have discussed the method for determining the carbon footprint of a product in Section 2.3.1.3.

Carbon management can be defined as management that identifies GHG emissions improvement opportunities from the production sites (and preferably from the entire life cycle of a product) and implements corrective measures to reduce GHG emissions. In order for the successful implementation of carbon management in a production site, GHG emissions should be quantified. Based on the quantified GHG emissions data, you can identify reduction opportunities from the production site.

3.4.2.1   Quantification of GHG Emission

There are several internationally recognized guidelines for the quantification of GHG emissions from production sites [11–13]. A GHG emissions quantification method is described in Section 2.3.1.3.

In order to quantify GHG emissions from a production site, the system boundary for data collection should first be defined. Here, the system boundary is defined by the production sites of the product and component parts. In other words, the system boundary is Gate to Gate which includes key parts (e.g., capacitors, resistors, printed wiring boards) for ERP and key unit manufacturing processes (e.g., extrusion, moulding). This includes collecting data such as GHG emissions from stationary combustion sources, mobile combustion sources, process emissions, as well as fugitive emissions. The GHG emissions data then becomes the GHG emissions database and can be used for GHG emissions accounting and certification.

3.4.2.2   GHG Emission Reduction in the Production Site

The analysis of quantified GHG emission data can reveal weak points in production sites in terms of the amount of GHG emissions. The contribution analysis is the method of choice here. In this method, GHG emissions from individual unit processes or facilities are divided by the sum of GHG emissions from the entire production site. Those unit processes or facilities contributing more than a certain percentage value (e.g. 1%) to the total, are identified as weak points. Corrective measures will then be applied to improve those weak points to reduce GHG emissions.

A common sense approach is often the best approach to reducing GHG emissions from production facilities. For instance, replacing existing low efficiency boilers and heat exchangers with higher efficiency boilers and heat exchangers can significantly reduce GHG emissions from the stationary emissions sources, as well as increasing efficiency (i.e. reducing operating costs).

However, a systematic approach is required when reducing GHG emissions associated with electricity consumption. In any systematic approach, the equipment requiring electricity for operation must be identified and the related electricity consumption determined (calculated or measured). Most equipment in production processes and utility supply facilities consume large amounts of electricity. Pumps and heaters, known to consume high amounts of electricity, can be replaced by ones with higher efficiency. In order to reduce electricity consumption systematically, however, a coordinated effort between process engineers and equipment suppliers are necessary from the equipment design stage onwards.

Table 3.2 summarizes major sources of GHG emissions and reduction measures from a production site.

Table 3.2 GHG emission sources and reduction measures from the production site

GHG Emission sources	Reduction measures
Stationary combustion (boiler, steam)	Increase combustion efficiency, waste heat recovery
Electricity consumption (production, utility, facilities, operation)	Design for energy savings during the facility design (equipment selection), optimum operation
Process emissions (chemicals)	Process optimization, installing abatement facilities, use of alternative chemicals

Alert

Carbon Management: For successful carbon management, it is advisable to expand the system boundary from the production site to all of the life cycle stages, including suppliers of parts, use, and end-of-life. Often GHG emissions from the production stage can be minor compared with the other life cycle stages, in particular for the use-intensive products such as ERP. In this respect, the carbon footprint of a product can be used to identify weak points of the product and, thus, corrective measures can be applied to reduce GHG emissions of a product over its entire life cycle.

3.4.3   Green Supply Chain and Regional/Local Aspects of Production

Current supply chain management is focused on ensuring the flow of a part containing no hazardous substances. For this purpose, green supply chain management systems, such as the Eco-partner and Green-partner, are in place along the part supply chain of most major electronics manufacturers.

Benefits of a green supply chain are not limited to the control of parts free of hazardous substances. The system can also be useful to implement Ecodesign. Information can be collected on material composition, weight, and the GHG emission data for each part in the green supply chain and the collected information can be used for Ecodesign. Based on the collected information of the part, a producer can evaluate its suppliers in terms of the GHG emissions potential and other environmental aspects of the part. The information and evaluation of results can be stored in the producer's parts database and can be used to communicate with suppliers.

Depending on the location of the production facility, different local and regional regulations may apply. Regulations such as air pollution control, wastewater discharge limits, and solid waste treatment methods differ from region to region and from country to country. Strict compliance with the applicable regulations is a must for any production facility.

Other major environmental aspects to consider in the production process include total nitrogen and phosphorus concentration in the wastewater, ozone depleting substance emissions in the air, GHG emissions, and toxic substances in the waste stream, among others. Nitrogen and phosphorus in wastewater can cause eutrophication, ozone depleting substances cause stratospheric ozone layer depletion, and GHG and toxic substances cause global warming and ecotoxicity, respectively.

Energy often comprises a significant portion of the product cost. In the case of the liquid crystal display (LCD) production plant, for instance, the electricity cost exceeds 1% of the total cost of the production of LCD. Thus, energy is not only an environmentally important factor, but also an important cost factor. For successful reduction in the energy intensity of a production process, an energy savings strategy must be adopted during the design of the production facilities. If this is not feasible, then optimization and improvement of the high energy consuming equipment should be a priority.

Checklist for Production Considerations – Section 3.4

Assessment questions	Answer	Comments	Recommended follow-up activities	Related sections
Have you considered four factors in production? Efficient use of resources; Efficient use of energy and utility; minimization of emissions; and Reduction of the use of hazardous materials	Yes ☺ No ☹	The goal of considering four factors in production is to minimize the consumption of the four factors during the production of the product	Perform analysis of the four production factors: resources, energy and utility, emissions, and hazardous materials	3.4.2

(continued)

Checklist for Production Considerations – Section 3.4 (continued)

Assessment questions	Answer	Comments	Recommended follow-up activities	Related sections
Have you analyzed the carbon footprint and implemented a carbon management system?	Yes ☺ No ☹	Reduction of the carbon footprint of a product and its production facilities is a must to remain competitive in the market	Develop carbon footprint, identify weak points of the carbon footprint, and improve the weak points	2.3.1.3, 4.4.4
Are you certified as a green supplier in the green supply chain?	Yes ☺ No ☹	Getting certified as a green supplier is a must to remain as a supplier to many finished product manufacturers	Establish and implement green supply chain management system	4.4.2
Do you comply with the local and regional environmental regulations on your production facilities?	Yes ☺ No ☹	Major environmental regulations on the production facilities include: Air pollution control, wastewater discharge limit, and solid waste treatment methods	Investigate different regulations in different localities and regions, and implement measures for the compliance with the regulations	1.2
Have you considered energy savings and optimized the energy use of the production facilities?	Yes ☺ No ☹	Energy savings are both environmental and cost factors for production facilities	Design production facilities for energy savings. Optimize and improve high energy consuming equipment	

3.5 Management Considerations

Environmentally conscious design will never be undertaken for environmental purposes alone. Management, by definition, has to balance costs, quality, and dependability together with the technical and environmental requirements. Different optima are possible for different kinds of products. In the final analysis, the products that sell best are those which meet customer requirements in the best way possible. Obviously, it is more profitable for a company to sell one million products, each of which reduces energy consumption by 10%, than selling a similar product which reduces energy consumption by 50%, but of which only a thousand can be sold because of design or appearance limitations!

3.5.1 What Approach to Take for Chosen Business Model

The decision for a business model [cf. Section 1.3], from the perspective of sustainability, must address, inter alia, the following:

– Integration of the application system of the product
– Integration of services

- Long-term planning
- Planning for reuse
- Life cycle thinking

As described in Section 1.3, many more factors should be investigated in order to select the most appropriate environmental solutions. But a change in business model from a company selling only hard goods to one with a service orientation, or one with a more life cycle based direction, opens many more opportunities for environmentally benign solutions. The change of Xerox from a manufacturer of copiers to a company selling everything in combination with documents "We are the document company!" enables leasing models instead of sales or take-back business models and allows recycling of used copiers and user friendly services.

Combined with the model chosen, questions arise about costs, quality, dependability, and technical features.

3.5.2 Costs

Many environmental costs associated with a product are usually not calculated when they are placed on the market and are used by the customer. Increasingly, environmental aspects or attributes of a product are becoming an important decision-making factor for customers, together with quality and price.

Costs should also be examined at all stages of the product life cycle. Costs cannot only be saved by changes in product design. Sooner or later, the impact of the total costs will return to the producer. With the application of the principle of *"producer responsibility"* by politicians, for example, the costs associated with recycling and take-back are being transferred to the producer. Some customers, such as those in the automobile industry, have taken the opportunity to load some of the external costs on their suppliers. Life cycle costs [14] are also a reason for increasing the cost of raw materials, particularly if the related mining activity creates environmental problems. Recycling is usually cheaper than mining. New ideas and innovations are needed to overcome these cost increases.

Examples for innovative cost reduction solutions

- Valuable packaging for IC can be cleaned and reused as new.
- Cardboard packaging can often represent 6–10% of the product's value. However, they can often be reduced to zero if the product can be sent without packaging!
- Maintenance and service costs can be reduced by environmentally friendly remote control.
- Software application can be substitutes for hardware or can help optimize processes.
- Software programs can be drivers of high energy consumption (e.g. by their battery loading commands).

More and more environmental regulations contribute to rising production costs or, as for carbon dioxide, certificates contribute to the costs of running plants. As a minimum, new investments in environmentally compatible technologies will need to be made in order to achieve or maintain competitive advantage. For other products, a cost increase could result because of regulations that require that hazardous materials be removed and substitutes found. A manufacturer has, through his Ecodesign process, the possibility to find cheaper alternative solutions or achieve some level of competitive advantage.

Promotion of environmentally improved products can be organized by governments. Examples include

- The "*toprunner*" approach in Japan.
- Ecolabels with energy classifications in Europe or in the US (with Energy Star).
- Direct intervention by governments (through taxes for cars, pay back, or prohibition of the worst technologies in the market such as the prohibition of incandescent light bulbs in some countries).

In this last example, old technologies will have to be replaced by more environmentally benign ones to avoid future cost increases.

Examples

"Toprunner" – Improvement by competition: The top performing product becomes standard after 2 years. Those consuming the highest energy have to be phased out. Prices can increase.

Energy classification is used for refrigerators in Europe – Class A++ and better in a scheme from A (best) to G (worst). The best products save roughly as much energy during their expected life as the difference in cost. The program is voluntary, but market pressures have pushed the average toward the better products; a similar effect is happening in Japan by the toprunner program.

Regulation: Some governments regulate the market in specific directions: Prohibition of incandescent light bulbs in Australia, Europe, and New Zealand has eliminated the use of less efficient lamps. The Ecodesign directive demands minimum requirements for all ERP in Europe. Differential taxes for cars with differing environmental impacts result in promotion of "cleaner" cars. Emission trading is another tool for reducing CO_2 emissions of plants.

Cost savings of between 25% and 50% are achievable by reusing "qualified-as-good-as-new" components [18], recycling of materials, or reselling of products.

Rising costs of energy and resources have to be integrated in the planning phase through the development of alternative or energy saving components.

A calculation of *activity based costing* instead of departmental costs is a much better way to identify the opportunities for product alternatives [15]. Environmental costs for end users are of increasing importance from a sales perspective. Even small cost differences may be important to the end user because of limited budgets. End users and purchasers are often not the same. The purchaser often only selects according to the lowest product price and not necessarily according to the total costs, and frequently without involvement of the end user. Getting the end user involved in the purchasing process may well lead to better long-term customer satisfaction.

3.5.3 Quality

Quality criteria should also extend to environmental requirements. During the introduction of RoHS, it could be seen that many quality departments were not interested in the regulation because they believed it related only to environmental issues. However, if the limit of a substance like lead is exceeded in a component, it could very well result in the total loss of that non-complying set of components and require the procurement of replacement components. As such, this kind of failure is a non-repairable failure. Complaints, therefore, will end up on the desks of the quality managers. To avoid these kinds of situations, environmental and quality managers will need to co-operate intensively and should integrate such complaints or failures in their early warning system.

Another new kind of quality problems occurs when environmental problems, together with the transfer of a production plant, are transposed to a developing country. Then problematic processes or chemicals may continue to be used. They have only disappeared from one location and are transferred to another! However, environmental rating agencies or the media may discover that a company continues to use special chemicals or processes somewhere in the world. When this happens, good news does not follow. The parent company might receive a poor environmental rating and any resulting public exposure could seriously reduce the company's market share.

Also, insurance companies may not be willing to pay for damages caused by substances which are already prohibited in many countries, if they are continued to be used in developing countries, even though these substances may not be restricted there! For instance, asbestos is still allowed in countries like China, India, and Russia. The illness resulting from exposure will be the same as it occurs in other countries. It will only be a matter of time before the lawyers begin to launch individual or class action law suits.

As quality people are accustomed to dealing with statistics, a new field for them will be the *risk assessment* and management of materials and components that are supplied to the company. The risk assessment can be combined with potential cost calculations on the basis that the worst case occurs – in which suppliers cannot supply! This is not the typical FMEA application but, in these types of circumstances, it will be necessary to inform the board of directors! Managers also need to see what a new law or regulation could cost as compared to the current state of the art. Examples are provided in Annexes 4 and 5. An early warning system can reduce the risk to nearly zero! But a good early warning system is only effective if (a) the information gathered worldwide is complete and distributed to all that need it in the company, and (b) measures are taken to deal with the risks.

Industry associations or manufacturers organizations (e.g. EIA, DigitalEurope, Orgalime, ZVEI) are good sources to help analyze trends in forthcoming legislation. Care also needs to be taken in dealing with import and export declarations. Company officials dealing with this subject need appropriate expertise and must be well informed.

It is always useful to integrate environmental requirements into any product development process. Many companies create their own checklists based on own experience (we refer here to the integration of IEC 62430 [Annex 9] into the development scheme).

Quality staff should look for milestones in product development to ensure that the required environmental attributes are, in fact, completely integrated into the product. Integration of the environmental attributes is not only necessary for components and products, they should also be integrated into capital projects, from design, to installation, to the final operation [Annex 2], and into services, maintenance, and software. Generally product, processes and management system have to be audited.

Suppliers should also meet the same kinds of requirements. Supplier audits often do not include questions about the integration of the environmental requirements into production processes related to supplied components or products. Much work can be saved if the suppliers have to fulfil at least some basic requirements, without which a relationship with the supplier may not be possible.

Alert

Generally speaking, customers assume that producers are prepared to, and actually do follow legislation and are a step ahead of any related trends, thereby ensuring that they will continue to operate in conformity to legal requirements. However, it is often the case that a customer is only interested in the lowest price of the supplied component. So his attitude might be "I do not want a change".

If these kinds of debates exist at the staff level, time is being wasted. In reality, it will be important to make the necessary changes early enough to accommodate the sales logistic needs of both the supplier and the purchasing company. Therefore such discussion should happen at the senior corporate level (i.e. the CEO level).

3.5.4 Technical

The technical challenges from the environmental side are manifold:

- "Innovation jumps" will be required for the product category sooner rather than later!
- Production processes that are out-of-date will have to be changed.
- Sometimes new technologies (e.g. for coatings or joints) may have to be applied without full knowledge of their long-term stability. How can we solve this problem?

Required environmental attributes are frequently an *innovation* driver. With old technologies, environmental improvement is usually very limited or perhaps even impossible. However, improvements in the range of 10–50% may be needed to fulfil world-wide demands for energy reduction, or for cost reductions in a market that has experienced dramatic increases in resource prices. The innovation potential for a washing machine working with water is rather low after 50 years of intensive development. Innovation is required with other washing liquids, materials, or technologies. [Examples: Washing with liquid carbon dioxide: [16] clothes that don´t become dirty by "Lotus effect": [17] Sanyo Japan].

Similar questions occur for old, out-dated industrial processes. They may have been leading-edge 50 years ago. But today, employee health issues could become costly if processes continue to rely on lead soldering, or if equipment cleaning processes continue to employ toxic chemicals. If alternative processes are evaluated, these costs or problems are often overlooked. Fortunately, alternative processes are available, the necessary conversion is not difficult, and the test costs are not very high. The biggest surprise – new technology processes can, in some cases, be much cheaper! Compression technology for joints, for example, can be cheaper than soldering.

A company should ask itself: Which kind of processes do we use? Is there a chance that, for example, the chromate coatings become prohibited? If so, what alternative is available? Do we sell the regulated motors by which we can, for example, save 50% of energy, or do we continue to sell old equipment? Customers will not accept poor technical attributes in the future.

The industrial processes of a company should be classified into:

(i) Those where there is high potential for restrictions on hazardous substances (rare or toxic) – in which case, changes are required immediately.
(ii) Those that use substances that could become cost drivers – in which case, alternative options should be seriously examined.
(iii) Those that use old processes but where improvement may be possible by changing ovens, drives, stand-by systems, better temperature management – in which case, some mid-term action is needed.
(iv) Those where the processes either need no immediate action or where the possibility of changes may be very limited.

The best way forward is to implement a *systematic process for environmental innovation*. One of the first issues in this kind of process is the age of a product or a technology when new product development starts. The predicted end of a product or technology should be estimated, as should the timing for the introduction of any new technology.

Alert

Managers might hear from their staff that no more improvement is possible or that 50% energy saving with this equipment can never be achieved. In reality, both objectives may well be achievable, but only with a new technology or by examining the full life cycle. In any case, the necessary investment must be organized by management.

3.5.5 Dependability

When new properties or behaviour is introduced into a product, customers will want assurances that the product has a high level of dependability and will perform over a considerable amount of time, particularly so when a price increase is involved.

The first question occurs when a customer calculates the energy saving over the product's lifetime. Perhaps the cost difference between an environmental premium product and an environmental poor product can be saved after some 10 years of product life. But if the normal guarantee for the new product is only 3 years, the investment might be at risk. The solution to this problem is an opportunity for producers and politicians!

In saturated markets, most policy targets can only be achieved if the existing products can be rather quickly substituted by those with better environmental qualities. However, no-one would willingly scrap a 5 year old refrigerator that still functions well and purchase a new one. The solution of this problem is again an opportunity for politicians and producers.

The concept is to include "qualified-as-good-as-new" components in new products. Many components have a life span that exceeds the normal expected life of the products in which they fit. In IEC 62 309 [18], the framework conditions are described under which no risk should occur for the customer. The environmental and the cost benefits can be very high. Nevertheless, the challenge is the "common wisdom" (albeit incorrect) that "reused products cannot be reliable". They can be reliable! The solution is to convince customers about the advantages of reuse and offer them the same quality checks and guarantees as for new products. Similarly, another "common wisdom" (again incorrect) is that environmentally compatible products cost too much and don't work as well as competing products. Companies could work together with associations and politicians to overcome the problems of the afore-mentioned "common wisdoms".

Alert

The long-term stability of new materials cannot always be tested sufficiently. In this case accelerated tests have to be applied. There remains a risk for some special applications but this should not prevent a company from applying such solutions. One could start testing as early as information about now legislation is available or when they determine that the product has, on average, a much shorter market life and the risk is much lower.

3.5.6 Management System

Management systems were installed as cost-saving measures and aimed at reducing failure rates and levels of work. A structured and systematic way to work is always superior to a chaotic way. Customers have more faith and trust in an enterprise that is well managed.

In this book, the authors have attempted to describe a number of environmental management activities and systems that are designed to achieve a more systematic approach to the implementation of environmental improvements and requirements. For Ecodesign, the inclusion of IEC 62430 into a product development scheme will often be sufficient (cf. Annex 9).

The easiest way to integrate environmental aspects into product design and development is to take – if available – the existing ISO 9001 system [20] and include aspects of IEC 62 430. In ISO 9001, the product development scheme is already included and, as such, the Ecodesign requirements easily fit. Many companies have also integrated Ecodesign in the ISO 14 001 system corresponding to the EC EMAS system. ISO 14 001 has its origin in the production and the way a plant has installed its industrial environmental protection management. Requirements for products are very rare in this standard and too general. A structure for Ecodesign is not available in ISO 14001. However, IEC 62430 fits into both standards. In Annex 9, we include an example for a possible structure for Ecodesign in ISO 14 001.

Indeed, in many enterprises *one* general management system containing the requirements of all necessary systems has already been installed. If required, certification can cover all of the management systems. "All in one" is naturally cheaper and integrates the different responsibility centres.

Alert

Not every management system must be certified! Only where customers require a certificate is it desirable. It is also possible to self-declare that an environmental system is in conformity with ISO 14001 or another standard.

It makes a lot of sense to integrate the requirements of ISO 9001(quality), energy management, OHSA 18000 (health & safety), and social responsibility. They all contain environmental requirements that should be dealt with together in an overall management system.

The next step – directed by software process requirements – seems to be the "process framework" [20]. This summarizes all of the repetitive processes in an enterprise in a structured hierarchy, with up to eight levels. If such a system becomes an international standard, it could potentially include all future management systems and their requirements in the corresponding software. Special extracts could allow for concentration on, for example, the environmental processes and requirements. Such a system could also be certified according to the management system standards if the elements of these standards were completely included in the "Business Process Framework".

In other chapters, we have mentioned several management systems, including:

- Energy management
- Chemical and material management
- Management of restricted substances
- Structured and planned innovation (ideas generation, patents, targets)
- Risk assessment (with suppliers, for the enterprise) and risk management
- Management of the whole life cycle including life cycle costing
- Management of reuse by a more general product planning
- Environmental supplier management

There is no need to complicate life by installing different and individual management systems for all of the different management requirements. For Ecodesign in the Ecodesign directive, Annex 1 provides a detailed description of the maximum requirements a company would have to meet for systematic implementation of Ecodesign (including documentation, target definition, alternative evaluation, and reporting for products to get the CE mark).

Management requirements are included in the Ecodesign directive and can be found in Annex 8. Requirements of the ERP CE-mark include environmental aspects in addition to the requirements of the European "Low Voltage Directive" (LVD) [21].

At the industry sector level in the EU, the mandated statistical reports require special reporting on a country by country basis. However, these figures do not necessarily reflect the reality for any one company. The Global Reporting Initiative [22] has developed a more ideal and standardized system in which many companies take part, but often not all have comparable world-wide data. Rating agencies like the Dow Jones Sustainability Group Index have developed more future-oriented reporting systems with figures that could influence the financial success of an enterprise. Reporting contents are changing every year and the ranking provides investors with information that could be used to influence decisions on stock purchase. More and more, financial success can directly relate to the degree to which companies manage environmental issues.

As previously mentioned, certification is not always necessary. Self-declaration could be sufficient, but customers must be willing to trust the declaration. On the other hand, *environmental audits* should be performed on suppliers and also internally. Together with the environmental audit, some companies sometimes forget to organize an annual *management review* with their management board. During this discussion, improvements identified by the audits should be decided upon and targets for the following year determined. It is important to describe the policy, the environmental principles and different organizational procedures in an "Environmental Principles" brochure. This type of information is required by ISO 14001 and allows interested parties to see what should be and what is done by the enterprise.

Checklist for Management Considerations – Section 3.5

Assessment questions	Answer	Comments	Recommended follow-up activities	Related sections
Have you chosen a sustainable business model?	Yes ☺ No ☹	It may be necessary to choose a model covering more than one generation of product, take-back and recycling, and service	Optimization, e.g., by including external partners, customers, competition, etc.	1.3, 3.1, 4.1, 4.2
Does your company want to become a market leader?	Yes ☺ No ☹	If yes, you must be best in environmental aspects and in management	Benchmarking	1.3, 2.5, 4.1, 4.5
Are your environmental aspects and strategic cost targets defined?	Yes ☺ No ☹	Future cost increases might depend on scarcity of energy, resources etc.	Make sure sustainability is part of company values	Annex
Does quality management include the fulfilment of environmental aspects?	Yes ☺ No ☹	Quality and environmental managers have to co-operate	Annual review	2.5, 4.5
Is your risk management planned and does it include environmental risks and potential costs?	Yes ☺ No ☹	Risks could come from new legislation, standards, changes in market conditions, changes in management systems	Regular update recommended	2.5, Annex 4
Are costs over the life cycle, or activity-based costs, integrated in the cost management for product development?	Yes ☺ No ☹	Many environmental projects would have a short payback period if they were calculated correctly	Check procedure	2.5, 4.5

(continued)

Checklist for Management Considerations – Section 3.5 (continued)

Assessment questions	Answer	Comments	Recommended follow-up activities	Related sections
Are Ecodesign directive requirements integrated into product design & development?	Yes ☺ No ☹	In a management system Annex IV or V of Ecodesign directive has to be integrated	Review for correctness	1.2, 2.5, 3.3, 4.3
Is a communication system planned to inform customers and the public about advantages of products and company?	Yes ☺ No ☹	Product declarations, green portfolio, future scenarios and targets	Test acceptance of information	2.2, 3.2, 4.2
Is a take back, reuse, and recycling system planned based on the acceptance of customers?	Yes ☺ No ☹	For refurbishment, dependability of the products or components and trust by customers are required	Build up environmental communication for your company and products	2.5, 3.3, 4.3, 4.4
Is there a list of all of the important environmental processes and their systematic management?	Yes ☺ No ☹	Management of hazardous substances, energy, take-back and recycling, innovation, suppliers, etc.	Identify regularly new processes	2.5, 4.5
Are technologies evaluated and systematically changed if required?	Yes ☺ No ☹	New processes need time to develop, therefore planning is necessary	Report about kind of processes and risks	2.5, 4.4
Is continuous improvement of the environmental strategy planned?	Yes ☺ No ☹	Basis: Global reduction targets, trends, competition	Observe market for environmental trends	2.5, 4.5

3.6 Integrated Strategy

By running through the checklists in this chapter, one can find one's own roadmap towards sustainability and identify where to go and why.

We have tried to identify the different considerations at the corporate, market, product, production, and finally management levels, and brought forth a range of ideas for possible implementation. All of these considerations should be combined into one picture in order to develop a company specific strategy.

The following decision matrix might be useful to develop the process strategy. It is advised to not only list strategies, but also assign responsibilities and define deadlines to achieve the identified improvements.

Level of consideration	Key question	Your own evaluation	Your own key strategy for improvement
Corporate	What code of conduct should the company follow?		
Market	How does the company want to position the company's products in the marketplace?		
Product	What does the environmental profile of the product look like?		
Production	How can the company's production processes improve environmentally, including issues related to energy and resource consumption, and hazardous substances?		
Management	How can the company ensure that the required (often strong) improvement takes place?		

Chapter 4
Action Plan Development

Following the situation analysis from Chapter 2 and the identified strategies in Chapter 3, it is now time to discuss concrete action to implement Ecodesign in the company. We know where the weak points are, and what the improvement strategies look like. It is now time to take action – again on corporate, market, product, production and management level. It is time to come up with an action plan.

4.1 Corporate Action Plan

A corporate action plan can help to introduce an environmental program in a systematic fashion. The environmental policy and vision will be the starting point for all employees and the public to understand the direction in which the company wants to move. Motivation and involvement of the staff is necessary to ensure that the action plan comes to life. As the subject matter can be very complex, training of selected employees and experts will be very useful. Distinct actions and communication will help create awareness in the public.

4.1.1 Develop Policy and Vision

ISO 14001 requires an environmental policy. The policy should include a focus on products and their environmentally compatible developments, the role of the stakeholders, customers and employees, and all of the related activities over the whole life cycle. As part of the development, all groups in the company should be consulted. This not only has the potential to add value but also serves to get considerable employee "buy-in".

The vision should be connected with the policy and provide a view as to where the company wants to be in the future. In implementing this vision, many companies develop a 5 or 10 year plan.

W. Wimmer et al., *ECODESIGN – The Competitive Advantage*,
Alliance for Global Sustainability Bookseries 18,
DOI 10.1007/978-90-481-9127-7_4, © Springer Science+Business Media B.V. 2010

4.1.2   Involve Staff

Environmental management systems require training of the employees. Internal newspapers, brochures, and information meetings can help to distribute necessary information.

A systematic uptake of information, and the related culture shift, is only guaranteed if environmental protection becomes part of every process and of daily work as has been the case with quality. This type of environmental training should be included as part of the orientation for new employees.

After management systems are implemented together with the PDCA (Plan-Do-Check-Act) cycle, product and production improvements can be planned. Annually, targets are contracted with all relevant departments and process owners, thereby, in essence, creating targets for every employee. The top down approach should start with the CEO who should set future oriented initiatives and targets. If the CEO has not yet agreed, it will be difficult for employees and customers to believe in the seriousness of any environmental engagement.

A "bottom-up" system for employee improvement proposals should integrate environmental subjects. The patent proposal system should have a focus on environmental issues such as energy saving. Environmental awards (e.g. to be annually granted) will attract the attention of many departments when environmental solutions are part of the success. With success stories in external publications, public attention can be drawn to the company and its progressive products and solutions.

Far too often, managers fail to recognize that voluntary engagement in the environmental affairs of the business can make employees proud and can enhance their reputation in the community. Good environmental behaviour by employees, such as organizing waste, running "take-back" initiatives in the plant, and finding ways to save energy, keeps environmental programs alive. The company is viewed positively by the public which, in turn, can draw attention and sympathy. As a result, employee engagement ends up, in effect, advertising for the company's "good" products.

Environmental engagement of the staff is also valuable for the supply chain and in dealing with customers. In many cases, for example when selling components or in projects with customers, the sales engineer can and should make proposals for environmental measures (such as energy saving) when they are not part of the tender requirements. Every customer will be delighted with value-added proposals; additional orders could well result and corporate image will be enhanced.

Educating retailers, such as dealers and retailers who do not belong to the company is often difficult, even though there may be a strong relationship. These retailers often "feel" that they have no environmental problem with the product. They want to compete only on the basis of cost. As they are usually not trained, and are not experts, they will not be able to properly inform their customers about any environmental aspects. It may require a lot of persuasion, direct information, brochures, and assistance to get these retailers interested and involved. This kind of customer engagement may well require the assistance of the management board, as resolution may be beyond the capability of a sales engineer. For him, facts are required and advantages can be directly given to the customer. As some costs will be incurred for implementing future programs, like the expensive take-back of capital goods, sales engineers should be sensitized as to where and how to include environmental aspects and costs in sales contracts.

Competent engineers should also inform customers about forthcoming legislation. If the information can be trusted, customers will come back to this supplier. Loyalty can be enhanced by providing training to customers on new legislation. The more environmentally compatible use of products (for example, Ford's Eco-Driving, and dosage recommendations for cleaning products) is another point where staff can have connections to customers. The environmental role of any company's staff cannot be completely planned and will be the result of personal engagement. Nonetheless, the structure and substance of this type of environmental customer service can be designed by management, who in turn should assist employees, especially those in the sales department, by providing necessary information (which could include examples of best practices, new products, and resource saving opportunities).

In-house trade unions can also be good ambassadors for the environmental initiatives and profile of a company. Their interest in the environmental activities of a company is most likely based on an understanding that high levels of environmental performance will help secure the future of the company. They should also be integrated in the company's training programmes.

4.1.3 Training

A variety of training courses will be very useful. A good example is the suite of courses and tools related to the ISO 9001 standard on Quality Management. Similar courses and tools would be helpful in fully implementing the ISO 14001 standard, particularly in reference to production.

As a first step, basic knowledge about Industrial Environmental Protection could be provided. As a second step, a course on Product Related Environmental Protection would be useful, followed by special seminars on Environmental Supply Management.

Which requirements should be provided to suppliers? What information is needed? The requirements for all suppliers should be integrated into contracts or as entry hurdles in internet based "supplier entrances". Procedures for information flow should be organized as is done for material declarations; a software tool should be offered which is compatible with standard software systems. It would make sense to include suppliers in environmental training courses.

Ecodesign tools and practical applications for design and development engineers could be another course, and it should not only include tools from quality, such as QFD and FMEA, but also checklists of best practice, evaluation of alternatives, calculation of life cycle costs, basic LCA, disassembly, and legal requirements. Strategic applications of design rules have to be combined with the CAD system.

Environmental engineers should be trained in Environmental Management (as per ISO 14001) and, if necessary, in Environmental Product Management (e.g. following IEC 62430). Auditor and quality training can be combined and should contain process, product, and system audits. Basic courses about auditing are offered by quality organisations and by some consultants.

For managers it is essential to become informed about existing and forthcoming environmental legislation, for example, on the product side, about chemicals (RoHS, REACH), take back (WEEE, batteries, packaging), and Ecodesign (ERP, IPP). Risk management, especially when anticipating forthcoming legislation, has to be part of any training. Additionally, it is very helpful to train managers as to how they should deal with the public and with media when problems arise. The loss of image and reputation as a result of poor environmental performance can be much more serious than with quality problems. Furthermore, government could shut down a plant immediately if environmental problems are not promptly resolved!

Alert

The Ecodesign directive is the first environmental directive for products that can kick you out of the market in the case of non-compliance. If the requirements are not met, the CE-mark cannot be affixed and, as a consequence, the product cannot be sold in the European Union.

The system of environmental standards (ISO and IEC, listed in Annex 11) should be included in any training and be offered to suppliers and contract managers as well as for internal application. The series of ISO 14000 standards do not deal just with management systems but also provide guidance for the different aspects and applications of Eco-labelling and LCA. For many customer requirements, and to be able to respond to questionnaires, it is very helpful to have standardized procedures in place and to know to which ones you could refer. Additionally, international standards at the concept or development stages can be influenced (or even initiated) in order to get the differing procedures of competitors or customers harmonized. Unfortunately, to date, not enough harmonization in these matters has been initiated. As a result, it can be a nightmare to respond to different kinds of enquiries from nearly every customer.

Trends for upcoming new standards and international legislation and regulations, should be regularly reported to management. In particular, product related requirements in standards can become part of new legislation. Like with the Ecodesign directive, where only a legislative framework is published, in the expectation that industry will organize the corresponding standards. Environmental elements are also being integrated into other standards, such as ISO 26000 for social responsibility, and there is discussion about harmonizing management systems. In China, standards are treated very much like laws.

Implementing new standards, and especially getting certification, can be expensive and frequently involves considerable amounts of bureaucracy. Prevention against too much of such bureaucracy is only possible if reaction starts in time. Industry is often too slow to react and, in the end, is forced to accept the certification procedures. Aside from the ISO, the IEC has also started and partially implemented new environmental standards (from IEC TC 111). A special energy management standard with many new elements is also on the way.

Alert

Because of complexity, a company should install at least one coordinating environmental expert. For daily work with Ecodesign, simplification is required. The co-ordinator could develop checklists and simple design tools (cf. Annex to this book). Workflow procedures should describe in detail what has to be done and, for example, the legal requirements.

4.1.4 Awareness Raising

Usually, the CEO officially starts the environmental programme and declares initiatives publicly (internal/external). In big companies, the fourth or fifth generations of such programmes is already underway. From the beginning, and with a lot of uncertain messages along the way, many programmes seem to mature and are now addressing real environmental problems. Positive awareness will come if the targets are ambitious, if the programme produces continuous improvements, and the published results can be trusted. NGOs will also evaluate the quality and degree of success of such a programme.

If risks associated with a new technology, new materials, or the use of hazardous substances are known, there will be rewards for companies that address possible problems in advance and offer solutions like for better health and safety. Corporate *image* will suffer if the company publicly maintains an unclear position. Like with Coltan (= Cobalt/Tantalum mineral) used for capacitors in cellular phones, where earnings were used for financing civil wars in Africa, no-one really knew its origins or where it was used when the press started to investigate. If a company is unable to provide timely and informed responses, then public confidence is lost. When this happens, a company now has to defend the product in order to continue its business and, in the case of "Coltan", agreements will need to be developed that ensure that "Coltan" or its components come only from "clean" countries and suppliers. Clearly, a proactive initiative by suppliers in the beginning will be much less expensive than the reactive approach described above and may avoid the not-at-all astonishing result of restrictions planned by government.

Alert

In the case

- Of disappearing acceptance.
- Of drafts of forthcoming legislation.
- That competitors offer better solutions.

Don't defend the critical material further. Start exploring alternatives.

After considerable bad media coverage in regard to lead soldering, the public could no longer be convinced that soldering with lead was a good solution! Even though it was not really shown that substitutes to lead solder could completely provide the same quality during the discussion of the forthcoming legislation, the legislative initiative could not be stopped. At this point in the legislative process, continuing to use the targeted technology becomes a risk for the company and for the customers. The technology is likely to be prohibited! The best a company can do then is to phase out the technology over time.

Public awareness is also raised by rating agencies such as the Dow Jones Sustainability Group Index (DJSGI). About one-third of the questions to participants are about environmental aspects. It can be shown that environmentally positive companies in these investigations have experienced better performance in the international stock exchanges.

Awareness can also be achieved if companies cooperate on environmental challenges and make joint public statements. If a group of companies would together declare that they do not use the most hazardous substances (e.g. asbestos) in their world-wide operations, then this message would create public awareness and also attract media attention. Such co-operation should include industry associations in order to avoid dissension and to raise confidence in the commitments. If these associations did nothing more than tell government, the public, and the customers that their members could do nothing more for environmental improvement, such information would be boring and is likely false! But associations can disclose a lot of positive information to the public. Associations can also promote new environmental advances or the availability of energy saving products from their members. In 2003, the US Engineering Associations, together with the US EPA, published the "Green Engineering Principles" [Annex 7 [1]] to show to the public the will of their members to undertake environmentally compatible work. Whoever is able to follow these principles can use it in their advertising. For the design of plants and factories, such a broad statement was completely new.

Another possibility for an association is to tell an astonished public that product solutions with energy saving for E&E industry are already available!

The progressive industry can already offer these solutions but industry needs individual assistance in (saturated) markets and governmental assistance to sell these products better [39].

Awareness in a market such as the E&E industry cannot be achieved by changing one component, or by eliminating one hazardous substance. Such advertisement creates no value-added for a customer and will be a target for competitors who will do everything to show other disadvantages of the competitor product. On the other hand, nearly all manufacturers in the E&E consumer products industry use the same technologies and have no real production depth. Therefore, alternative components or substitute materials are usually not available! Changes often have to be co-ordinated with competitors or even with the corresponding manufacturer organization or they will be impossible to implement!

The use of environmental product declarations is a new way to explain not only the environmental advantages of a product to customers, NGOs, government, or the general public but also to do advertising. For the ISO 14020 series of standards refer to Section 1.2.3. However, unless standardized across the product category, comparisons between products may be difficult because of differences in test methods, reporting, and life cycle boundary conditions. In addition, customers may demand expensive independent verification. With the Ecodesign directive, ecolabels might, at least in part, integrate environmental aspects into the CE mark.

As environment is of increasing importance to customers, they want to have guarantees that all substance restrictions are observed. They will have more requirements for a supplier, and they will, more and more, want to audit the supplier processes. In the end, certificates may also demanded for environmentally sound activities wherever they take place in the world: hazardous substances in China, rain forest protection in Brazil, etc. Such certificates cost money.

It is possible that, with long, worry-free relationships, customers will trust their suppliers. However, to maintain this trust, suppliers will need to stay abreast of developments that may impact on the environmental performance of both their company and the products they produce. Environmental reputations need to be properly cared for and nurtured.

Environmental reports and reporting (including internal ones) and annually published results show the public what has been achieved to improve the environmental performance and situation of a company. More requirements than just those associated with environmental protection, such as social responsibility and sustainability, have been added to the reporting requirements and the term "environment" often disappears from the titles of the reports. Some companies publish "sustainability" reports. Others integrate environmental aspects into their annual business reports. Environmental properties influence living conditions world-wide, and the management of a company has to prove that they are responsible, not only for naked figures but also the company's impact on environmental and social conditions. Environmentally more compatible products, designed and produced for developing countries, such as cooking stoves that work with local oil resources, have a low price, and are assembled in the developing country [2]. They raise public awareness because they show that western companies are willing to offer products for those in developing countries.

As environmental properties have become a competitive edge, awareness is also guaranteed by the activities of competitors if they offer essential environmental improvements. As a nomination in the DJSGI becomes important for investors, and as only the first few (often only three) in a category are mentioned, the fight for the best places becomes harder.

Positive environmental results of some companies are recognized by the public and some organizations grant environmental achievement awards to excellent managers [in Germany by BAUM [3] an environmental organisation of industrial companies; it grants annual environmental award for excellent environmental managers] or for excellent solutions. Companies can also create their own awards, but have to be very careful that the prize committee will be believed to be independent. Publication of the winners creates a positive image.

Awareness is also guaranteed if the CEO is integrated in the audit and review process and sets targets such as reducing the carbon footprint of the company. Companies that buy a big area in the rain forest to preserve nature and want, by that, to neutralize the environmental impact they create as a result of their own production, do something positive for the environment. However, it can easily be claimed that such an activity is only trying to buy favour, rather than actually making their business more environmentally compatible.

As such, a better approach is to redesign their own product and improve their own manufacturing and supply chain to reduce their carbon footprint.

Well informed customers will certainly understand the difference in the approach.

Checklist for Corporate Action Plan – Sections 4.1

Assessment questions	Answer	Comments	Recommended follow-up activities	Related sections
Is a plan available by which activities and processes, policy, and vision will be implemented?	Yes ☺ No ☹	Describe processes, implement mid term program for transition	Update regularly	2.1, 3.1, 4.5
Is management and staff included in an information and training programme?	Yes ☺ No ☹	Top-down, starting with CEO, bottom-up with ideas and proposals	Update programme annually	2.5, 4.5
Is there a system installed to involve and assist the employees to convert environmental proposals and projects?	Yes ☺ No ☹	Besides verbal assistance, budgets have to be agreed upon if benefit is acceptable	Report improvements	4.5

(continued)

Checklist for Corporate Action Plan – (continued)

Assessment questions	Answer	Comments	Recommended follow-up activities	Related sections
Is systematic training developed with basic information, Ecodesign, industrial environmental protection and required information for all kinds experts?	Yes ☺ No ☹	Since much know-how is required, such training should be in-house! Besides environmental experts, engineers, purchasing people, sales need special information	Use multiplicators to spread the information	3.5, 4.5
Are suppliers also involved in training?	Yes ☺ No ☹	Know-how is not always available to suppliers	Continuing	2.5, 3.4, 4.5
Are required tools available?	Yes ☺ No ☹	Software, LCA, checklists have to be offered to those people who require them	Cooperate, e.g. with universities, if own expertise is not sufficient	2.3, 2.4, 3.5
Are there awareness programs for the public and for the company?	Yes ☺ No ☹	CEO can start a programme with big promotion!	Regular activities also in the Press are necessary	4.5
Are environmental product declarations developed?	Yes ☺ No ☹	These fact sheets are required for all external activities	Publish step-by-step EPDs for all products, also plants	2.2, 3.2, 3.5
Is the carbon footprint of the product calculated and communicated?	Yes ☺ No ☹	Inform yourself which standard of the carbon footprint is expected from your customer	Calculate and communicate the carbon footprint of your product. Follow example in Section 5.4	Example in 5.4
Is there an environmental reporting programme installed to update the reports?	Yes ☺ No ☹	Gather required figures, install software, include suppliers, production information system	Project for several years required to install, e.g. software, over all processes	3.5, 4.5
Is there an environmental auditing process in place?	Yes ☺ No ☹	Production, product and system audits are all necessary. Suppliers need to be audited. Auditors have to be trained. Quality and environmental engineers need to cooperate!	Also include recyclers and international affiliations	2.5, 3.5, 4.5

4.2 Marketing Action Plan

4.2.1 Marketing

A recent article by Roger Cowe of Context, a CSR Consultancy, pointed out that it was high time to wake up the marketing department. The message wasn't about some new product about to be introduced. It wasn't even about some improvement to an existing product. It was all about challenging experts in the communications and marketing fields to direct their skills toward sustainability and corporate responsibility issues.

It has long been acknowledged that consumers, on average, are not sufficiently interested in, or knowledgeable about, sustainable development and environmental issues. The same is also likely to be true for most new products introduced into the marketplace. The Sony Walkman is a prime example. When about to be released, market research indicated that people didn't understand the concept and thus weren't interested in it. This didn't cause the marketers to back down. Yet when it comes to the myriad of environmental issues, most notably climate change in recent years, where are the marketers? Are they waiting for consumers to demonstrate more concern? That certainly wasn't the approach taken by Sony. Their marketers got busy doing what they do so well and introduced the word Walkman into the English language.

The process of marketing can lead consumer attitudes and thereby create market advantage for those willing to make environmental investments. And we're beginning to see some hints of that. If the trend continues, the companies that don't keep up with consumers' environmental attitudes will fall behind. If successful, their businesses will achieve market advantage from their environmental investment.

In most businesses, marketing involves understanding the needs and motivations of clients and then finding ways to satisfy these needs and motivations. The fundamental proposition is that we (as society at large) have not "positioned" the "problem", taking into account what most effectively motivates human beings and companies to take action – their own immediate self-interest.

As mentioned earlier, effective marketing begins with an examination of the problem or opportunity, continues with a review of how the opportunity might be exploited, examines the motivations and needs of the client, and then designs (through ECODESIGN) and positions the product, concept, or service accordingly.

4.2.2 Environmental Responses

Dealing with the various types of environmental problems can bring long-term economic benefits as a result of technology and process changes and better long-term sustainability of resources. There are a number of different but complementary ways to deal with environmental problems. Command-and-control mechanisms are perhaps most familiar but others are also available. Economic mechanisms – such as financial incentives or disincentives offered by governments – are increasingly being examined and applied. Social marketing, the use of moral persuasion, is also often used by governments to "correct" or encourage societal behaviour.

Other market-based approaches – such as systems of tradable permits and environmental labelling, which lever the spending power of individuals and organizations to reward environmental investment through market activity – are also available.

Each approach has both positive and negative aspects and may be more appropriate for certain types of environmental problems. Regulation, for example, is best applied when there is no direct incentive for companies or individuals to undertake the environmental improvement required.

Social marketing campaigns, on the other hand, have usually adopted one of three key themes. One theme is the message to others that the environmentally damaging action should be stopped because of the dramatic environmental consequences that will occur. In other words, environmental problems are often framed as a "direct issue" (see earlier classification) with potentially dire consequences. Front and centre in a poster produced by the Pesticide Action Network in the United Kingdom, for example, is the assertion that "Long-term effects of regular exposure to pesticides often cause chronic illness, including cancer, reproductive and neurological effects. The World Health Organization believes that pesticides cause 772,000 new cases of disease each year". Clearly, people are encouraged to take note!

A second recurrent theme is that people should change their environment-related behaviour because it is somehow "the right thing to do". An example of this is the "What Would Jesus Drive" campaign launched by the Evangelical Environmental Network in 2003 – there is also the suggestion that our action in terms of the vehicle we drive should reflect better behaviour: "Transportation is now a moral choice ...". When the advertisement's authors say that it is about more than vehicles, "it's about values", they are certainly playing to peoples' needs to "do the right thing".

And finally, the third theme involves promoting environmental action because of the economic benefit it will bring. Do this environmental action, the message proceeds, because it will save you money. However, as already suggested in our reference to energy-efficiency improvements above, we feel that this approach is more accurately characterized as an "economic mechanism". Yes, it is often part of social messaging, but it is an economic message being transmitted.

4.2.2.1 What Is Missing

What we conclude, however, is that relatively little attention has been paid to the way in which a "marketing approach" could be applied to environmental challenges.

As can be seen, industrialized societies have developed a range of responses with regard to how environmental challenges should be met. The use of political, economic, and social mechanisms in isolation and combination has served to improve the state of environmental quality from what it otherwise would have been.

What is equally clear, however is that, collectively, these responses have not been enough. The recently-released Millennium Ecosystem Assessment was yet another reminder that more needs to be done as societies respond to our growing environmental challenges. Furthermore, in an April 21, 2005 article in *The Economist* entitled "Rescuing Environmentalism", the following is quoted:

> Mandate, Regulate, Litigate. That has been the green mantra. And it explains the world's top-down, command-and-control approach to environmental policy making. Slowly, this is changing. Yesterday's failed hopes, today's heavy costs and tomorrow's demanding ambitions have been driving public policy quietly toward market-based approaches.

What strikes us as missing from the traditional portfolio of responses, and what might be alluded to in *The Economist* article, is a "marketing approach" – what we call "environmental marketing". In other words, let's add to the discussion about environmental responses by thinking more systematically about how a marketing approach might offer new insights, ideas, and strategies for developing effective responses. This should be particularly of value when applied to those environmental challenges where some benefit for taking action can be identified (i.e. most indirect and resource-management challenges).

While the connection we make among motivation, information, and environmental marketing is, we argue, innovative, we certainly accept that others seem to be arriving at a similar destination, if for other reasons. In other words, there are signs that environmental marketing has to "go beyond" the green consumer and that it has to move away from its traditional messages of "catastrophe" and "altruism". We see such evidence in both the academic literature and actual marketing campaigns. For example, recent scholarly writings have begun to identify the importance of exploring consumers' "self-esteem", considering, for example, the extent to which a product purchase affects the purchaser's image in his or her social group [4]. Similarly, other analysts have reflected upon the "emotional brand benefits" of particular product purchases; these include, for example, "auto expression benefits through the socially visible consumption of green brands" [5]. Evidence of such change is also appearing in the field.

Environmental marketing should be aiming at the majority of the population (see previous Figure on Notional Population Distribution) with its messaging. In other words, it should be recognizing that "self-gratification" is what might be motivating the vast majority of consumers. Once the appropriate "motivator" is identified, it then becomes the task of graphic artists and creative writers to translate their message to meet the information needs of the prospective clients.

4.2.3 Information Needs

If environment can be positioned effectively against individual or company needs, and thus motives, the next key step is appropriately packaging the relevant information. While the previous section argued that effective marketing should be aimed at meeting self-gratification needs, this does not mean that environmental information should be entirely avoided. Rather, that the orientation of this information should not be based on fear or altruism.

Information on the environmental aspects of the product, activity, service, or facility needs to be tailored to the receiving audience. Every audience is unique. Each has its own capacity, interest, and ability to receive, understand, and then act on information. Therefore, environmental information must be carefully prepared and designed with an appreciation of the target audience's values, understanding, and level of attention available.

Figure 4.1, again from a TerraChoice Environmental Marketing Inc. presentation on "Marketing Pollution Prevention in North America", serves to demonstrate that there are many environmental "dialects" out there. The figure identifies the different environmental issues that are included in purchasing decisions in terms of frequency (often, occasionally, or never). Of particular interest are the pairs of circled items. The red circles identify that Energy Conservation is always either "Often" or "Occasionally" required, while some 15% "Never" require a Climate Change criteria. This is of interest because there is a very high correlation between energy efficiency and climate change. However, climate change is an environmental issue while energy efficiency is considered as an economic issue. Similarly, the blue circles highlight the purchase requirements for "VOCs" and "Human Health" which also correlate very highly. However, VOC's (an environmental issue) are of considerably lower interest to purchasers than Human Health Issues.

The conclusion from this graphic is that communication of environmental information must be tailored to the issues of importance to the buyer.

Values are also particularly important in that individuals and companies have an array of environmental values that range (as per Roper) from True-Blue Green to Basic Brown. Those closer to the True-Blue Greens are likely to want and act on direct environmental information, while those closer to the Basic Browns will want information more related to their values (indirect environmental information).

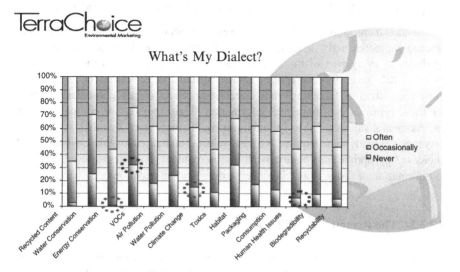

Fig. 4.1 Environmental Dialects

Of course, a different conclusion can be reached about the distribution of "greenness" of the population. Rather than thinking of "greens" and "browns" and shades in between, one could also conclude that there are those with more "eco-centric" values and those with more "anthropocentric" values. This does not mean that those closer to the "anthropocentric" end of the spectrum are opposed to environmental issues, but rather that other values are of higher priority.

4.2.3.1   Audiences at the Eco-centric End of the Value Spectrum

Many consumers with an interest in environmental issues, and if motivated, are likely to want limited, focused, relevant, and credible environmental information on the product or service being considered. Thus, comprehensive information based on the life cycle stages (i.e. extraction, transportation, manufacturing, use and disposal) of the product or service is not likely to be of much interest or use to this kind of consumer.

- To satisfy the "limited and focused" requirements, the consumer will likely want to see a few descriptive words and/or a simple label or seal of approval.
- To satisfy the "relevancy" requirement, the consumer will want comfort that the environmental claims and benefits are not offset by something much worse (in terms of environmental performance) in a different part of the life cycle (e.g. low energy use, but with highly toxic waste). The consumer will also want to have confidence that the claim is better for the product or service in question than competing products. In other words, if a product is claimed to be non-toxic, then in order for it be relevant, at least some competing products must be more toxic.
- To satisfy the "credibility" requirement, the believability factor comes into play and this can most often be met by having an independent verification undertaken.

On the other hand, large companies with an environmental interest, and well-informed consumers, increasingly demand extensive information on all of the environmental attributes in every part of the supply chain. Companies, such as those with ISO 14001 certification, those preparing annual sustainable development reports, and those committed to "greening" their purchasing processes, are particularly hungry for those types of environmental information. Those persons or companies with environmental values are already "environmentally motivated". In other words, little persuasion is likely to be required for these selected audiences to make environmental choices. Appropriate information is all that is required. For our purposes, we refer to information of this sort as "*direct*" environmental information.

4.2.3.2   The Remaining Audiences

For individuals or companies who do not value environmental issues in the same way as those at the "eco-centric" end of the street, information must be tailored to the kinds of things that are important to them. In many cases, environmental information can be translated into a "dialect" that fits these values ("*indirect*" *environmental information*). For example, climate change translates easily into clean air for children, cost saving through energy efficiency, and energy security for the country. Environmental concerns associated with pesticide and chemical use can be translated into a reduction in incidences of asthma and allergies, thus reducing health costs.

Similarly, information that positions the product, service, or behaviour as cool, sexy, improving image and status, etc., can also be effective.

4.2.4 Communication Tools

The "Market Recognition Framework©" (Fig. 4.2) was developed as a guiding structure for packaging and presenting both direct environmental information (for greener audiences) or indirect environmental information (for remaining audiences). The foundation elements for the Market Recognition Framework (MRF) relate directly to the need for relevant and credible information and form the base of the MRF pyramid. The other elements of the MRF describe the different approaches that can be taken to provide environmental information that is appropriate for the targeted audience.

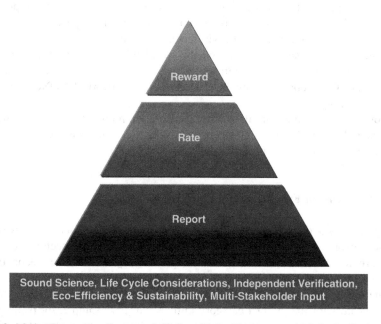

Fig. 4.2 Market Recognition Framework (© TerraChoice Environmental Marketing Inc.)

4.2.4.1 Reports on Environmental Performance

In order to provide credible and relevant reports on the environmental performance of a product, activity, service, or facility, ideally the foundation activities of sound science and life cycle considerations should be followed. Specifically, claims such as "better", "less toxic", "low energy", and "less polluting", would not be allowed in this type of report.

Environmental performance reports do not judge a product, service or facility to be good, bad, or indifferent as these are comparative assessments. The reports simply provide relevant, credible, and useful information. Comparisons are left to the recipient of the report who can use the information to assess the performance of the product or service relative to competing products or services.

The main environmental attributes should be provided in a format that has been standardized for that particular sector, as should the testing methods and protocols (i.e. the way the measurements are taken). Even something as common as recycled content can be measured in different ways. For example, the recycled content of paper is normally reported as a percentage. However, is it a percentage of the final weight of the paper or the fibre content? Does it refer to only post-consumer fibre, or does it include pre-consumer waste such as post and pre-industrial materials and manufacturing scrap? Relying only on a claim of "recycled content" doesn't allow the consumer to compare that aspect of environmental performance between products unless it is clear how each product represents its recycled content.

Furthermore, these kinds of environmental performance reports should be as objective as possible with no social valuation of, or relative weighting between the environmental attributes being reported. This way, the recipient of the report can compare information for similar products or services in an "apples to apples" kind of way and based on her/his own environmental priorities.

A number of organizations around the world have developed Environmental Product Declaration (EPD) programs of the "reporting" variety and most are members of the Global Environmental Declarations Network (GEDNet website is www.gednet.org).

However, environmental information is also often reported on single attributes. For example, recycled content, energy consumption, and non-toxic ingredients reports are often found on product labels.

4.2.4.2 Rating of Environmental Performance

Using relevant and credible information reports on a product, activity, service, or facility sector, a relative rating system can be developed for that sector. The rating can be represented in a five star system, in a gold, silver, bronze structure, or simply as a percentage score.

The main environmental factors that differentiate between products are used as key indicators for the product category in question. For example, for paper products, key indicators or attributes such as energy burden, recycled content, forest management, water consumption, and a range of pollution metrics could be used as the basis for a rating system for paper products.

In order to reduce each of the key indicators to a common scoring system, so that a single score can be calculated, three steps are required:

• Determine the score for each key indicator relative to industry performance ranges.
• Determine and apply weights, or priorities, to each of the key indicator scores.
• Combine the weighted scores into a final score which can then be assigned a relative rating.

The weighting process can be undertaken in a number of ways, and relies on assigning relative priorities to each key indicator. Often, this process is undertaken by an expert panel or by a policy discussion and analysis of relevant considerations, such as the spatial and temporal impact of the key indicator, its severity, and its relevance to the sector in question. Whatever process is chosen, each key indicator is assigned a relative priority, or weight, which is then used to modify the relative scores of the key indicators.

The final rating can then be used to compare one product, service, or facility against others. Rating systems are easy to quickly understand. Hotels and restaurants have had such systems (but not for environmental performance) for many years. More recently, TerraChoice has developed environmental performance rating systems for hotels, golf courses, and marina facilities, which allow consumers a quick and easy way to understand how the facility rates from an environmental perspective. Environmental rating systems can similarly be developed for any product, activity, service, or facility.

4.2.4.3 Rewarding Environmental Performance

With relevant and credible information on a product, activity, service, or facility sector, and a relative rating system in place, the "best in class" can be easily determined and "rewarded" with an environmental label or logo. This "best in class" normally refers to the upper 10–20% of the products, services, or facilities in each sector.

A number of environmental leadership programs exist around the world and the Global Ecolabelling Network [6] has been established to allow for an exchange of information and to further international coordination and cooperation.

In Canada, Environment Canada's Environmental Choice Program (ECP) has been in operation and awarding its EcoLogo™ since 1988. The ECP was the second such program in the world (preceded only by the German Blue Angel Program, 1977). Today, more than 35 similar programmes have been developed in Europe, Asia, and North and South America.

To use environmental information in the marketing of a product, activity, service, or facility, the target audience becomes a key determinant in the approach chosen. Environmental Reports tend to fit a well-informed commercial audience (or a very well-informed consumer), while rating and reward systems tend to better fit the average consumer (or a commercial or government operation with limited time or interest to digest substantial information).

The Market Recognition Framework© operates as a "vertically integrated" structure to provide the most appropriate information for the audience in question. In this context, "vertically integrated" means that a credible and relevant reporting system cannot be developed without the foundation elements being observed. A credible and relevant rating system cannot be developed unless an information-reporting system is in place. And, finally, a reward-based system cannot be developed unless some form of rating system is in place.

Some companies combine reporting and rating systems to assist them in making purchasing choices. For example, the Portland, Oregon, organization Metafore is working with a range of large multinational companies to help them in their paper purchasing. They have developed an Environmental Paper Assessment Tool, the EPAT^{SM} which relies on a standardized Environmental *Report* on the range of life cycle based impacts for pulp and paper products, and allows buyers to apply their own environmental priorities in order to determine the relative environmental *rating* of competing paper products.

Regardless of the target audience or market, and assuming that the marketer has been able to identify the appropriate motivators, the final key to marketing lies in the language of communication. Not the marketer's language, but rather the recipient's language. Hence, the need to identify whether direct or indirect environmental information will work best for the audience in question.

4.2.5 Conclusions

Environmental marketing has historically and typically focused its attention on that segment of society that is already "there" from an environmental awareness, concern, and action perspective. This focus has probably limited the uptake of the environmental message to those 20% already so inclined. Ironically, it could be argued that such environmental marketing may, in many ways, have been wasted on those already committed (that is, "preaching to the choir").

The proposition is to move away from the *values*-based approach of past environmental marketing (that targets the green 20%) to broader *needs* or *motives*-based marketing (that targets the entire marketplace).

"Do the right thing" and environmental "fear" are not the best motivators to engage the majority of society. These messages have been heard too frequently during the last 40 years. The fact that air and water quality are, on balance in North America, Europe, and many parts of Asia, better today makes the message not only stale but no longer believable. Furthermore, few people in North America actually "live" in these motivation spaces.

Environmental marketing needs to evolve in such a way that it becomes aligned with what actually motivates the majority of society – that is, various forms of self-gratification or self-interest (the achievement of supportive families and communities, quality friendships, respect of peers, etc.). Message themes must therefore coincide with what actually motivates the bulk of the population. Status, image, cost savings, coolness, children's health are but examples of the kinds of positioning that resonate well.

How the information is presented is also critical. Too much technical jargon is not helpful to the bulk of consumers. However, the informed commercial buyer or citizen may want detailed and technical environmental information.

Non-environmental language can often be used to convey environmental messages, as strong correlations exist between many environmental issues and those of more immediate interest to the bulk of society (e.g. toxics correlates with health, global warming correlates with energy conservation and cost savings, indoor air quality correlates with worker productivity).

Alert

Can your environmental message be criticized for not being fully honest? The six principles of environmental marketing (see Section 3.2) are good guidelines. To recap, they are:

(1) Be cautious about highlighting one environmental attribute, while ignoring other potentially more significant environmental attributes.
(2) Ensure that any claims are, as a minimum, verifiable. Often, the best approach is to have an independent party verify claims.
(3) Avoid any claim that is non-specific or vague. A term such as "natural" or "chemical-free" can be both true and false depending on interpretation and context.
(4) Ensure that claims are relevant. Claiming that your home ink-jet printer is free of DDT is technically correct but there are no printers made with DDT.
(5) Similar to the first principle, avoid promoting the environmental aspects of a product that is, by definition, harmful to the environment, just a bit less so than competing products. A company's cigarettes will not be considered green just because they use less packaging, and have lower levels of tar and nicotine.
(6) Be truthful.

Checklist for Marketing Action Plan – Section 4.2

Assessment questions	Answer	Comments	Recommended follow-up activities	Related sections
Have you established the objective for your environmental communication?	Yes ☺ No ☹	This could include increased sales, introduction of a new product, or even raise corporate image	Review corporate objectives and environmental objectives	3.2, 4.2

(continued)

Checklist for Marketing Action Plan (continued)

Assessment questions	Answer	Comments	Recommended follow-up activities	Related sections
Have target audiences for environmental marketing been identified?	Yes ☺ No ☹	Audiences include consumers as well as professional buyers		3.2, 4.2
Have you assessed the knowledge of your target audience and their primary motivators?	Yes ☺ No ☹	Knowledge levels vary greatly, from limited and narrow to broad and even technically well-informed	Customer survey	3.2, 4.2
Have you determined the key messages?	Yes ☺ No ☹	This can relate to safety, health, cost as well as, of course, environment	Tailor the message to the target audience	3.2, 4.2
Have you decided how the messages should be transmitted?	Yes ☺ No ☹	A variety of options exist, from third-party logos to self-declared statements	Tailor the approach to what is credible for your customers	3.2, 4.2

4.3   Product Design Action Plan

What could be done by the product development department? How can an engineer in product development start working on the environmental improvement of a product? What needs to be done for the different product types outlined in Section 3.3?

From Section 3.3 – product considerations – we know that different Ecodesign strategies are required for the different environmental profiles of products. It has been shown how to improve use-intensive products and raw material, or manufacturing intensive products.

Certainly there are different strategies for different product types. Nevertheless, we can identify basic principles for improvements which apply to any kind of product.

4.3.1   Basic Principles

For actual design changes, we can identify three basic principles which can be applied to most environmental design tasks. These basic principles are:

- Reduction
- Substitution
- Avoidance

"Reduction", as a basic principle, refers to using less material in a product, to reducing the energy consumption in the use stage of a product, and any other reduction in resource consumption along the life cycle of the product. The basic principle of reduction usually does not fundamentally change the product but improves it on the basis of the existing product concept. This could result in reduced product weight, reduced energy consumption during use, or even reduced production waste.

Example – Action to reduce energy consumption (Fig. 4.3):

Minimize energy consumption at use stage by increasing efficiency of product

Active (current using) products may consume more resources at use stage than was required for their manufacture. Usually, energy consumption for the operation of the product is a crucial factor. The European energy label indicates different classes of energy consumption for various products (washing machines, refrigerators, freezers...) is to support the customer in his purchase decision. While, on the one hand, low energy consumption is used as a positive feature in advertising, suppliers also try to push sales emphasizing the high power input of certain products (kitchen appliances, vacuum cleaners...). This is used to underline the outstanding power of the product, although it does not say much about the quality of a product. New paths should be considered to communicate high performance at low consumption levels (e.g. for vacuum cleaners by indicating cleaning power, expressed in max. pressure difference instead of power input).

Fig. 4.3 Checklist example [16]

"Substitution", the second basic principle, refers to the exchange or replacement of a component, but can also refer to substituting one certain function of the product for another. As such, the actual product concept may undergo fundamental change. There are also other forms of substitution which do not necessarily influence the entire product concept, such as substituting a certain problematic material with another less harmful one or the substitution of a certain auxiliary process material in the production line. These tasks are sometimes challenging, but do not necessarily change the entire product concept. In most cases the basic principle of "substitution" requires more effort than the basic principle of "reduction".

"Avoidance", as a basic principle, refers to omitting a certain key element or function of the product. This, in most cases, is connected with a totally new product concept. For the example of energy consumption, one can understand that reducing energy consumption is manageable within a certain product concept. Even substituting one source of energy for another can be eventually done with only minor changes to the product concept. However, avoiding the use of energy altogether will certainly require a totally new product. Therefore, this third basic principle of "avoidance" can be considered the most challenging, but also the most promising in terms of the potential for significant improvement in the environmental performance of a product.

Example – Action to avoid material consumption (Fig. 4.4).

Example – Avoid Production Waste (Fig. 4.5).

The three basic principles can now be combined with the product characteristics laid out in Section 3.3. Table 4.1 shows the application of the three principles to the use-intensive and raw material/manufacture-intensive products.

Avoid and/or minimize waste at use stage

Type and quantity of waste generated during the use stage of a product may considerably influence the total environmental balance of the product. Strictly speaking. only active (consumption-intensive) products consume energy or materials at their use stage. Considering the whole life cycle of the product, the consumption of these resources usually dominates the overall environmental impact of the product. Therefore, the prevention of waste at use stage is very important with this type of product. the example shows the use of rechargeable accumulators instead of conventional batteries.

Fig. 4.4 Checklist example [16]

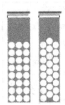

Avoid waste and emissions in the production process

The goal of each production process consists in the transformation of raw materials into products. Thus, process waste may be considered an indicator for inefficient use of materials. Apart from the environmental impact caused by the disposal of waste the consumption of raw materials extracted from the environment has to be taken into account. In may cases, the procurement of raw materials that are transformed into waste in inefficient production processes, is a decisive cost factor. Avoiding this type of waste not only reduces the cost of disposal but also purchase costs for raw materials.

Fig. 4.5 Checklist example [16]

Table 4.1 Product improvement principles and actions for use-intensive and raw material/manufacturing-intensive products

Product characteristics	Reason	Principle	Actions (examples)
Use intensive	Energy consumption	Reduce	❏ Minimize energy consumption by increasing efficiency of product
			❏ Reduce stand-by consumption
		Substitute	❏ Choose different principle of function
			❏ Make possible use of renewable energy resources at use stage
		Avoid	❏ Avoid environmentally harmful abuse of product
			❏ Avoid use of batteries
	Material consumption	Reduce	❏ Design product for minimum consumption of process materials
			❏ Concentrate wear on replaceable components of product
			❏ Make signs of wear easily visible
		Substitute	❏ Design for use of process materials from renewable raw materials
		Avoid	❏ Avoid and/or minimize waste at use stage
			❏ Close cycles for process materials needed at use stage
			❏ Provide for incentives for and possibility of collecting waste from use stage

(continued)

Table 4.1 (continued)

Product characteristics	Reason	Principle	Actions (examples)
Raw material/ manufacturing intensive	Material	Reduce	❑ Realize simple principle of functioning
			❑ Reduce number of parts and components
			❑ Integrate functions
			❑ Aim at optimum strength
			❑ Reduce packaging
		Substitute	❑ Use of materials with a view to their environmental performance
			❑ Use of recycled materials
			❑ Use single material components
			❑ Reduce number of different materials
			❑ Prefer materials from renewable raw materials
			❑ Prefer recyclable materials
		Avoid	❑ Avoid the use of new material by reusing parts and components
			❑ Avoid or reduce the use of toxic materials and components
			❑ Avoid inseparable composite materials
			❑ Avoid raw materials, components of problematic origin
	Process	Reduce	❑ Use energy efficient production technologies
			❑ Reduce energy consumption by optimum process design
			❑ Minimize overall energy consumption of production site
		Substitute	❑ Preferably use renewable energy resources
			❑ Preferably use process materials from renewable raw materials
			❑ Use environmentally acceptable auxiliary and process materials
		Avoid	❑ Avoid production waste and recycle process materials whenever possible
			❑ Avoid environmentally hazardous production technologies *and materials*

Example – Action to Integrate Functions (Fig. 4.6):

Reduce material input by integration of functions

A critical review of the structural design with a view to the functions needed may result in a considerable simplification by integration of fuctions. Combining several functions in one component does not only reduce material input but also facilitates assembly and disassembly as there are fewer connecting parts.

Fig. 4.6 Checklist example [16]

The example actions in the box above are taken from the ECODESIGN PILOT – a checklist based online tool, which is available in ten different languages under www.ecodesign.at/pilot.

In Annex 2a examples of improvement rules for products are given and in Annex 2b examples for plants and projects.

A systematic approach to the above mentioned checklists and product improvement strategies can be found online under: www.ecodesign.at/pilot (Fig. 4.7).

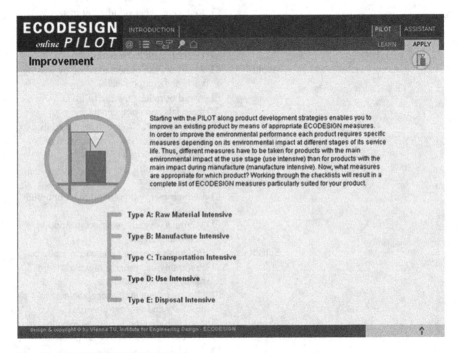

Fig. 4.7 ECODESIGN PILOT [16]

The process of achieving product improvements can be divided into three steps:

Step 1

Interpretation of the existing product's environmental profile – understanding how the most significant environmental impacts are caused and identification of the most critical materials, functions, parts, or components. Additionally, based on the disassembly analysis and a potential new product structure, some discussion should be held on the disassembly barriers.

4.3.2   Example TV-Set

As an example, the environmental profile of a TV set is presented in Fig. 4.8.

From the environmental profile one can understand that the TV-set is a use-intensive product with the main environmental impacts caused during use. The next question is what contributes to the impact in this use stage – certainly the operation of the device, the energy stand-by features of the device, and the batteries used for the remote control. That is all if we assume that no repairs are done and, as such, no spare parts are needed at the use stage. If one now examines the three causes of the environmental impact the main part – about-two thirds are caused by operation, about one third is caused by stand-by energy consumption, while only a minor impact comes from the battery use.

Assuming a use scenario of 4 h of watching TV and 20 h of stand-by per day, the total energy consumption for only the stand-by feature over 7 years (the expected lifetime of the TV) is 300 kWh. Consequently, some 600 kWh are needed for watching TV.

Following Table 4.1, and going through the list of actions, one can

- Minimize energy consumption by increasing the product's efficiency
- Reduce stand-by consumption
- Choose a different principle of function

The first two actions may result in a different technology for the screen like an energy efficient LCD-panel. The last action, regarding stand-by, will require some creative ideas of how to get rid of this unnecessary energy consumption or at least produce some significant reductions. Targets, or objectives, should be set because their limits help to determine the potential solution.

When applied across the production run of any one product, the potential for improvement can be quite large. Let's examine the example of televisions. Assuming that the manufacturer produces approximately one million TV sets of

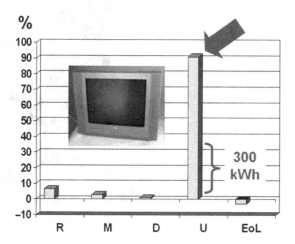

Fig. 4.8   Environmental profile of a TV set

this model, then the energy consumption caused only from stand-by over the lifetime (7 years) of the TV results in 300 GWh which is roughly one-third of the annual energy production of a large river water power plant.

With that example, and the demonstrated multiplier effect across a production run, it is easy to understand that even little consumption factors should be examined for reduction.

When designing plants, factories, or production facilities, the approach is similar to the energy-intensive product.

4.3.3 Example Office Chair

If one imagines a raw-material-intensive product like an office chair, then different improvement strategies are needed (Fig. 4.9). From Table 4.1, the concept of material reduction translates into

- Realizing the simple principle of functioning
- Reducing the number of parts and components
- Integrating functions, etc.

This requires careful examination of the product structure – preferably through a disassembly analysis. The idea is to understand the product's functions – what is needed and why? – but also to learn about the different disassembly levels.

Disassembly is often not in focus when doing product design and development. Although a smart disassembly concept cannot only lead to improvements in assembly

Fig. 4.9 Office chair [17]

and therefore in strong cost reductions and also in a well thought through product structure, but can also allow access to the most valuable parts and components for later reuse.

Most of the time one finds too many disassembly levels. They can be simplified by

- Defining new functional (integrating) units
- Reducing the myriad of material types and parts

Structural optimization should be a guiding principle for the development engineer to bring some order to the design, and as well help to strongly reduce costs.

In the case of an office chair, the product structure is important not only for recycling of materials – mono-type materials would be ideal – but also for the aspect of upgradability. If, due to a smart disassembly concept, the surface elements (e.g. seat, cushions, etc.) of the chair can be changed easily, then the product, in essence, creates new business opportunities. For example, the manufacturer of the easy-to-disassemble and upgradable office chair is no longer only a supplier, but can become a business partner, since he can also offer after-sales upgrades, such as changing the surface from green textile to black leather. This way, the customer gets the unique opportunity to upgrade his office by changing its look and feel without necessarily changing all of the furniture which usually has a useful life of 15 years. This is already a strong selling feature for those in the office furniture business.

Step 2

In the next step, the product design team works on new ideas and solutions following the product improvement actions laid out in Table 4.1. The targets, and possibly a new product structure, should be achieved. In the case of the TV set example, this may be a more energy efficient screen and significantly reduced stand-by energy consumption.

Although the focus should be on energy consumption, the structure of any new product has to be well defined. Here, it would be an ideal target to reduce the total number of components, which make up the TV, to only 5 or 6, and ensure that they are easy to disassemble. Examples could include one environmentally compatible plastic for the housing, with no metal inlays, and a different one for the cables. Reduction of materials in this way also creates benefits in logistics, purchasing, assembly time in manufacturing, and material consumption.

Improved energy efficiency would logically be part of improvements in the electronic component(s) but could be combined with software solutions for avoiding or reducing the energy consumed in the stand-by mode.

Step 3

Once the new product design is finished, the improvements found need to be assessed. How much of the planned environmental improvement have been realized. This is important to know since there are always trade-offs.

Let us assume that one finds a much more energy efficient TV screen through the development of a new type of display technology. This would certainly be a major improvement that would require a significantly new design. But then new questions arise:

- What types of materials are required?
- How much energy is needed to produce the new TV screen?
- Are there any other environmental problems involved with the new screen?
- Who supplies the new screen?
- Does the supplier have an environmental management in place and are the environmental data provided reliable?

So, for the assessment in this step, the entire life cycle perspective should be used to determine to what degree this new technology and its requirements fit with the other materials, parts and solutions. The improvement must be recalculated from this higher-level perspective. Also, new potential environmental risks may well arise and will need to be evaluated. As a result, there may be some iteration in the design process until a fully acceptable solution is chosen.

After the final improvements are quantified and are sufficient to convince customers and the public about the positive environmental improvements achieved, the marketing department should be contacted to develop an appropriate environmental marketing strategy in order to inform consumers and the public about the product's improved environmental performance (see Fig. 4.10). A more detailed plan on how to run the Ecodesign improvement process has been described in a 2004 guidance document [7].

The "product improvement" process takes place after identifying the Ecodesign tasks and is comprised of improving the use features and actual design work, all of which is target and outcome oriented and related to the reduction of a product's environmental impact.

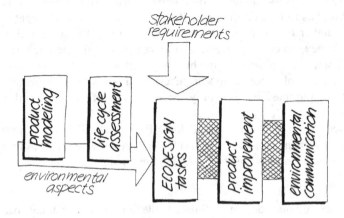

Fig. 4.10 Ecodesign roadmap [7]

Fig. 4.11 Product development process [7]

This design process is clearly not about environmental issues alone. Issues such as quality, cost, and safety also need to be considered. As such, the result of the design process is always a compromise, delivering the best achievable result under the limitations of the product specifications that guide the entire design process.

The early stage of the product development process (see Fig. 3.5) is very important. If there are no clear specifications about the environmental performance of the product like energy consumption or recycling rate, then there will be no such performance in the redesigned or new product.

When defining the product specifications for a new product, it is most useful to have an already calculated environmental profile of the previous product in order to identify on what to focus at the very first stages in the design process.

Coming back to the TV example, the importance of addressing the energy consumption is apparent. But how does one set the new specifications for the new design? Should the new TV be 10% better in energy performance or more than 50%?

Certainly the answer to this question varies from product to product. However, there are a few ways to help establish the specifications. The first way is to examine the physical minimum of energy required for a certain function. This is easy for applications that provide heating or cooling. Here, one can calculate the energy needed to boil a litre of water or to cool a particular space, like with a fridge. Then one can compare the energy efficiency of the company's product to the physical minimum.

Another way is to compare the company's product with the best available technology on the market and compare the differences in efficiency.

If it is not easy to calculate or define a minimum of energy consumption, one could benchmark with other similar products on the market and identify possible reduction targets as product specifications for the redesign process. Benchmarking is always possible and not limited to energy consumption. It can be also done with other environmental improvement aspects.

All the action laid out in this Section can only be realized and integrated into product improvements if critical success factors are considered. These critical success factors are

- Top management supports the Ecodesign project and its objectives – including allocation of resources (budget, time input, etc.).
- An adequate organizational structure is established, a project team is appointed, and responsibilities of experts and decision makers are defined.
- Ecodesign principles are integrated at the beginning of product development.
- Challenging targets for improvements are set in the product specifications.
- Multidisciplinary teams are included in the evaluation aspects and in the development of new ideas.
- "Out of the box" thinking is encouraged in order that really new ideas are explored at all levels of potential improvements.

All in all, the success of integrating Ecodesign ideas into new products depends on

- Understanding the location of the environmental weak points of a product
- The capability to translate the weak points into specific improvement tasks
- Availability of the right tools to fulfil the sometimes new tasks of environmentally improving a product; and last but not least
- A top management decision to go for it

So it's now the time for the managers to go out and motivate the product development teams to do the right things.

Checklist for Product Action Plan – Section 4.3

Assessment questions	Answer	Comments	Recommended follow-up activities	Related sections
Do you know what is causing the main environmental impact of the product?	Yes ☺ No ☹	Identify the life cycle stage and key environmental parameter	Look for alternative technologies	3.3
Do you know what product parts and components are most relevant from an environmental perspective?	Yes ☺ No ☹	Identify them using the environmental profile		3.3
Do you know what are the most relevant (new) product design specifications with which to begin the product development?	Yes ☺ No ☹	Perform benchmarking, calculate efficiencies, look at best available technologies	Set further improvement targets, although if no benchmark is available	3.2, 3.3, 3.5

(continued)

Checklist for Product Action Plan (continued)

Assessment questions	Answer	Comments	Recommended follow-up activities	Related sections
Do you provide adequate tools and support to your design team for their decision making processes?	Yes ☺ No ☹	Provide design checklists ·	Develop the lists by applying the knowledge of the team	3.3, 4.5, Annex
Did your team achieve significant environmental improvements?	Yes ☺ No ☹ ·	Perform environmental assessment of the achievements and evaluate them	Set new targets	3.5, 4.5
Are suppliers involved – do they know the new requirements?	Yes ☺ No ☹	Inform them and make sure they deliver according to the new specifications	Assist suppliers if necessary	4.1, 4.4
Is environmental communication and marketing department informed about the achievements?	Yes ☺ No ☹	Support the development of environmental communication about the new product	Increase the number of environmental product declarations	4.2

4.4 Production Action Plan

As discussed in Section 3.4, there are four major factors to consider in the production consideration. They are the minimization of the consumption of resources, energy and utility, emissions, and hazardous materials. Thus, the goal of the production action plan is to ensure efficient use of resources, efficient use of energy and utility, minimization of emissions, and reduction of the use of hazardous materials in the operational boundary or production site. All these activities are linked directly to the operation of the production site where reduction of the production cost and environmental impact are important goals. In particular, demand for the disclosure of the GHG emissions from the operational site and the product entail a heavy burden on the companies.

4.4.1 Production Information System

Customers are increasingly concerned about the use of hazardous chemicals in the production process and which do not remain in the product itself.

Table 4.2 Shifting Trend in customers' requests for environmental information on supplied parts

Phase	Period	Scope	Content
I	1995–1999	Cleaning agents used in the unit processes	Use of ozone layer depleting substances
II	2000–2001	Restricted use substances in the parts/products	Regulated heavy metals and others
III	2002	Type and content of major materials and substances in the processes and products (including certificate)	If regulated substances are used, show their content
IV	2003 to present	Type and content of all materials and substances in the processes and products (including certificates)	Information disclosure on the composition of the entire part/product on a mass basis, by application and composition and by mass balance of all chemicals used during production

[8]

Accordingly, the production action plan and its related information system should address all of the chemicals used in all of the various processes. A supplier management system should also have requirements related to hazardous chemicals.

Today, most finished product manufacturers or producers require their suppliers to supply environmental information on all supplied parts. The type of information can take various forms, ranging from environmental management system certificates to declarations confirming that no hazardous substances are used in production or in the part itself. There is a definite and distinctive trend in the scope and content of the environmental information requested. Table 4.2 shows the shift in the environmental information requirements by the manufacturers on finished products, from basic information needs on specific chemicals used in the production processes to a wide spectrum of hazardous substances in the part and/or product.

Table 4.2 shows that producers are continually expanding the scope and depth of the environmental information requirements on parts from suppliers. This is a clear indication that product related environmental concern is attracting more and more attention from regulators around the world, and stringent regulations aimed at the environmental aspects of products are emerging. Of particular interest in Table 4.2 is the focus on hazardous substances in the evolving environmental information requests. This coincides with the hazardous substance regulations of products that are typified by the RoHS and RoHS-like regulations in the EU and other parts of the world. One can expect that this trend will soon progress to include energy efficiency and Ecodesign related information such as LCA data of components.

Table 4.3 includes the kind of information typically requested by producers that is specific to hazardous substances. The information ranges from environmental management systems to product specific attributes, such as the level of assurance that lead-free soldering was used. Information on hazardous substances in the

Table 4.3 Specific information requested by producers from suppliers on hazardous substances

Information type	Information specific to the hazardous substance control
EMS certificate	• Meeting the criteria set by the buyer on RoHS regulated and other hazardous substances controlled by the producer
	• Prerequisite to the green supply chain management certification
Non use certificate	• Declaration that supplied parts do not contain RoHS regulated and other hazardous substances
	• Proof that the regulated substances are not present in materials used for supplied parts
Raw material composition	• Preparation in accordance with the material declaration form provided by the producer
	• Necessary to trace hazardous substances in the materials in case of non compliance
Analysis results of hazardous substances	• Proof that supplied parts meet maximum concentration values
	• Preparation in accordance with buyer requirements
	• Certificate issued by a certified laboratory
	• Expiration period: 2 years

[8]

Table 4.4 Application of RoHS regulated substances in the E&E industry (class I)

Substance	Major parts and processes	Applications
Pb	Lead soldering, lead battery, rubber, plastics, glass, CRT, ceramic condenser, electrode, etc.	Rubber solidifier, colouring agent, paint, lubricants, plasticizer, battery material, etc.
Cd	Electrical contact, Ni-Cd battery, plating, PVC coating of a power cable, etc.	Colouring agent, rust proof treatment, stabilizer, plating material, etc.
Hg	Electrode, mercury battery, DC battery, lamps, fluorescent tube, back light unit of LCD, etc.	Rust proof agent, high efficiency light emitting device, colouring agent, electrical contact materials, etc.
Cr^{6+}	Chromate steel plating, battery, color filter, rustproof chromate treatment, etc.	Colouring agent, paint, ink, catalyst, plating, rust proof treatment, etc.
PBBs*, PBDEs*	Printed wiring boards, power cable coating, connector, switch, fuse, switchboard, etc.	Flame retardants

*PBB = Polybrominated biphenyls, PBDE = Polybrominated diphenylethers [8]

supplied parts/products is the common denominator in Table 4.3. This again reflects the current need for disclosure of information on the use of hazardous substances in products.

Table 4.4 lists the six RoHS regulated substances and the major parts in the E&E industry where they have typically been used. In the E&E industry, the RoHS regulated substances are termed class I substances. Meanwhile, there are other substances controlled by the manufacturers because of concern that they will become regulated in the near future, or will be targeted by environmental NGOs or consumer organizations. These substances are termed class II substances. Most E&E producers ask for information about both class I and class II substances. They often also ask for information on the material composition of parts, also known as material declarations [9].

Environmental information can have many purposes. One of the major uses is to communicate the environmental aspects of the part or product to the marketplace. For this purpose, it is important to offer LCA data, especially for materials. Several manufacturing organizations have already published such data (e.g. PlasticsEurope for different plastics). Thus, for customers requiring LCA data of components, such as the automobile industry, the desired data are often available. However, databases including those from governmental organizations, are not always up to date.

These types of data are also needed in any communication initiatives, especially environmental product declarations, such as those described by ISO 14025. The other important application is in the exchange of environmental information on parts and materials, the use and end-of-life stage of a product among the participants in the supply chain. This is an indispensable facet of the production of ecoproducts.

Figure 4.12 shows the information exchange for Ecodesign along the supply chain.

In some cases, there are systems for the supply of these data, such as with the automobile industry's IMDS system. IEC PAS 61906 is widely compatible with this system.

An overall information system with the input and output of chemicals for all production processes would be the final step in developing a specific production information system. This would allow for complete LCA's to be developed, assist in identifying reduction opportunities, and help answer questions from the public and rating agencies, like the DJSGI (Dow Jones Sustainability Group Index).

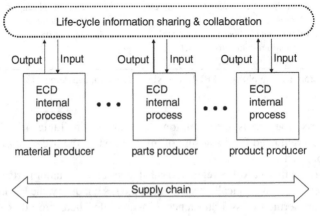

Fig. 4.12 Information exchange along the supply chain for ECD Processes [10]

4.4.2 How to Build a Green Supply Chain Management System

In order to produce an ecoproduct, the supply chain has to be green. Setting up and maintaining green supply chains has long been the practice in the E&E industry and it has been widespread since the late 1990s after the enactment of the RoHS regulations in EU member states.

The green supply chain aims at ensuring that parts and materials supplied to the producer by its suppliers meet the environmental requirements of the producer. The green supply chain is basically parts and materials control system implemented by the producer. Figure 4.13 shows the parts and materials control system for the production of a product by the producer.

Depending on the level of risk of a supplier's parts, inspection frequency can vary. Those with high risk are checked once a week, while with lower risk, perhaps monthly. Table 4.5 shows the applicable parts and their inspection frequency in accordance with the defined level of risk of those parts.

A supplier can set up its own parts' control process in order to meet the requirements of the producer which means that the supplier establishes a process that identifies the composition of raw materials of the part and generates detailed analytical data for the RoHS regulated (class I) substances in the parts. Normally, producers help suppliers to build a parts/material control system as shown in Fig. 4.14 [8]. With a properly functioning supplier's parts control system, the producer can ensure that he meets the environmental requirements related to the product.

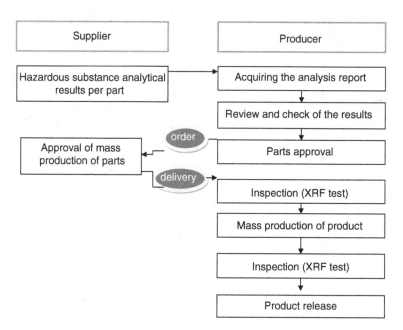

Fig. 4.13 Manufacturer's parts and material control system [8]

Table 4.5 Parts inspection frequency

Risk level	Applicable parts	Frequency
I	Seven part groups of high risk (extrusion, painting, plating, power cord, rubber, wire, plating)	Once/week
	Parts with history of violating the RoHS regulated substance control limit	
II	All parts excluding the risk level I parts	Once/month

[8]

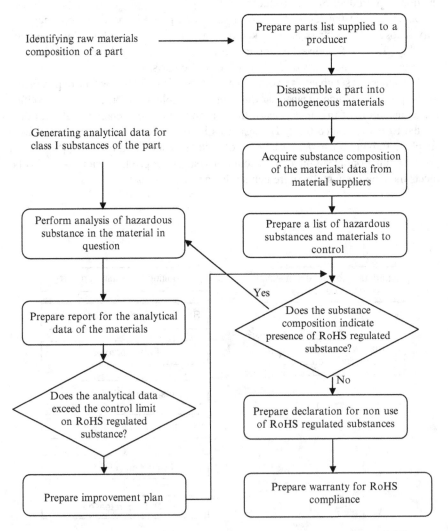

Fig. 4.14 Supplier's parts control process (Hazardous substance management system) [8]

4.4.3   Operation and Implementation

Once a green supply chain management system is in place, the producer can operate and implement the system. The typical mode of operation/implementation is through certification of the supplier's parts/material control systems. Such programs as the Green Supplier Programme or Green Partner Programme are examples of the operation and implementation of green supply management systems.

The producer audits the system based on documentary evidence and results of on-site visits. The system includes the Environmental Management System (EMS), and the materials and parts control system. Table 4.6 shows the audit items and weighted score of each item that is assessed against three levels: good, average, and poor. Scores in Table 4.6 are examples common in most of the E&E industry. A passing score, in general, is 80 points out of 100. However, compulsory items enclosed by parentheses ( ) must score more than 80%. Upon successful completion of the audit, the supplier is awarded with the green supplier certificate.

4.4.4   Carbon Management in the Production Site

Measuring the consumption of resources including materials and energies, utilities, and emissions is the major operational activities in the production site. Since these are all also sources of the GHG emissions, every effort must be made to develop an accurate inventory of all these consumptions as well as finding reduction opportunities. Therefore, a production action plan considers ways and means of collecting the consumption data of materials, energy, utilities, and emissions and then finding the reduction opportunities.

Table 4.6 Green supplier certification system assessment

Assessment item		
Major category	Sub category	Score ( )
EMS; presumed to meet the EMS requirement if certified by ISO 14001 (30)	Policy and strategy	10 (4)
	Internal audit	10 (4)
	Training	5
	Information sharing	5 (2)
Hazardous substance management system (including RoHS substances) (40)	Improvement plan	11 (8)
	Control of non-conforming parts	11 (5)
	Control of changes in parts	8 (7)
	Control of its suppliers	10 (5)
Material and parts control (30)	Incoming parts inspection	12 (6)
	Material and process control	7 (3)
	Outgoing parts control	11 (6)
Total		100 (50)

[8]

The collected consumption data can be converted into GHG emissions data using the GHG emissions calculation formula and the emission factors. Thus, the production action plan is identical to the carbon management plan. The methods and practices for the collection and calculation of the GHG emissions, and the identification of reduction opportunities of the carbon footprint of a product, apply here.

Alert

Carbon Management at the Company Level: Two areas where major cost effective opportunities can be found are energy efficiency and the use of renewable materials. The other two areas that offer cost effective improvement opportunities are transport and waste.

GHG emissions from the product system can be used to find reduction opportunities over all of the product's life cycle stages. In practice, however, most manufacturers cannot control the suppliers or users of their products. Furthermore, manufacturers are liable to report and reduce the GHG emissions from their production sites. Thus, reduction in GHG emissions can only be realistically made within the operational boundary or the production site. The collection of the GHG emissions data from the production site and identification of the emission sources to reduce its GHG emissions is referred to as carbon management of production. The production site carbon management can be implemented as shown below. All the data collection and calculation of the GHG emissions discussed in Section 2.3.1.3 apply here. The only difference here is that the production site is the system boundary instead of the entire life cycle of the product.

A typical procedure for the carbon management of a production site is shown in Fig. 4.15. It helps to identify the necessary steps and corresponding action items.

Steps and action items listed in Fig. 4.15 have been applied to the management of carbon in many companies. Below is an example application of the procedure to an electronics part producer.

Example

The system boundary was set up by defining the production site of the part. The identified GHG emission sources are electricity, process emissions, and other minor emission sources such as fugitive emissions. The GHG emission calculation method, similar to the one described in this book, was chosen for the calculation. The GHG inventory data by compiling and calculating the GHG emissions data shows that 49% of the GHG emissions come from the production processes, 48% from the use of electricity, and the remaining 3% from miscellaneous sources including non point sources.

It was found that the GHG emissions from many of the production processes, mostly GHG chemicals such as SF_6 and perfluoro carbons, cannot be reduced because of the nature of the processes. Thus, the only viable option is to manage energy consumption including electricity, utilities, and other minor ones such as lighting.

Steps	Action items
Identifying and determining emission sources	Setting up system boundary of the production site, listing emission sources, preparing a process tree, and identifying the current status of the measuring equipment
Selection of the calculation method	Preparing GHG calculation guide, conformity assessment procedure, and data collection format
Developing inventory	Compiling the GHG emissions data, developing the GHG emissions inventory, and preparing the inventory report.
Evaluating potential reduction amount and setting up the reduction target	Evaluating the potential amount of the GHG emissions reduction, assigning priority for reduction, and running simulation of the reduction target
Managing reduction plan	Developing a reduction plan for major emission sources (e.g. electricity, SF_6) and reporting the reduction performance
Evaluating reduction performance	Managing performance evaluation indicators, compiling statistics of the reduction performance, and analysis of the statistics

Fig. 4.15 Carbon management procedure for a production site

Table 4.7 Potential amount of carbon reduction and the assigned priority

Category			Consumption (MWh)	Ratio (%)	Priority
Production	Electricity	Main facilities	428.1	28	1
		Auxiliary facilities	187.0	12	
	Utilities	Compressed air	442.3	29	1
		Ultra pure water	112.5	7	2
		Air distributor	73.4	5	4
		Cooling water	23.2	2	5
Support for the production		Cooling equipment	142.9	9	3
		Air circulation fans	82.5	5	5
		Lighting	26.9	2	5
		Wastewater treatment facility	7.5	0.5	5
		Pumps	10.3	1	5
Total			1,536.6	100	

Table 4.7 shows the evaluation results of the potential reduction in GHG emissions and the relative priorities for the reductions.

Based on the potential carbon reduction and the assigned priority, the reduction plan for major emission sources (e.g. electricity, SF_6) should be developed. The reduction action plans for the major sources with a priority of 1 and 2 should be given high priority. First, develop reduction targets. Second, develop reduction action plans as shown in the example box. Lastly, evaluate the reduction performance using the performance indicators such as percent reduction of the energy consumption, etc.

Example

Reduction action plans for the reduction of the energy consumption in the production site of an electronics part manufacturer: Measure electricity consumption of the production facilities and diagnose the consumption: compare the electricity consumption of all production facilities between measured values and the value of the equipment specification. Any deviation should be diagnosed to rectify the deviation or excessive consumption.

Reduce energy consumption for the generation of the compressed air: diagnose optimum consumption of the compressed air for each production facility and rectify any deviation from the optimum consumption, renovate the production facilities to enable operation at lower pressure, separate high compressed air generating facility from the normal ones.

Reduce energy consumption by reducing the use of the compressed air in the production process: replace production processes that require large amounts of compressed air, such as Air Knife and Air Curtain operations, with production processes requiring lower amounts of compressed air.

Reduce the use of ultra pure water in the production process: reduce it by increasing the recirculation of the used ultra pure water back to the production processes.

Don't be surprised to find high reduction potentials of as much as 30% or more! Install measuring equipment where you can identify energy consumption drivers. Usually there is only one metering system available at the entrance of the factory!

Checklist for Production Action Plan – Section 4.4

Assessment questions	Answer	Comments	Recommended follow-up activities	Related sections
Are you aware of what specific hazardous substance information is required by producers?	Yes ☺ No ☹	Information on the product specific attributes on hazardous substances, such as reliability of lead-free soldering, can be drawn from the environmental management system	Identify what specific information is asked by the producer and prepare for it	4.4.1

(continued)

Checklist for Production Action Plan (continued)

Assessment questions	Answer	Comments	Recommended follow-up activities	Related sections
Have you set up a parts control process or hazardous substance management system?	Yes ☺ No ☹	You as a supplier should establish a process that identifies the composition of the raw materials of the part and generates detailed analytical data for the RoHS regulated (class I) substances in the parts	Establish a part control process within your production process	3.4.3, 4.4.2
Have you integrated production and product information into one process?	Yes ☺ No ☹	Such a system is very valuable, as it is flexible – it can be used to answer questions from the public and to control the system (and costs)	Establish an integrated system for reporting, cost evaluation and LCA	4.4.5
Have you passed an audit of your Environmental Management System (EMS), and materials and parts control system?	Yes ☺ No ☹	Passing the audit is prerequisite to being a green supplier; however, continual improvement through operation and implementation will be required as well	Pay attention to the compulsory audit items to meet the requirements	4.4.3
Have you set up a carbon management procedure in your production site?	Yes ☺ No ☹	Carbon management in the production site is identical to the management of energy, materials, and utilities used in the production site	Establish the carbon management procedure similar to the one in Fig. 4.15	2.3.1.3
Do you routinely update the carbon reduction target and evaluate the carbon reduction performance?	Yes ☺ No ☹	Proactive approach in setting up more stringent carbon reduction target ahead of the market can give competitive advantage to the company in the long run	Develop performance indicators for the evaluation of the carbon reduction performance of the production site and make the evaluation an integral part of the production-site management	2.3.1.3

4.5   Management Action Plan

Ecodesign should become a systematic approach in every company. Therefore, the step-by-step *implementation* of an Ecodesign management system should be the first action if there is no other system available. A framework for the related requirements can be taken from the Ecodesign directive, Annexes IV or V.

As many product specific requirements (of which only some are environmental) should be integrated, an *extension of the system* should be planned, including safety, fire protection, technical requirements from product standards, Environment, Health and Safety (EH&S), and social responsibility or sustainability requirements from customers. As a result, a wide, complete but unsorted collection of all potential requirements to derive a new product profile will be available. Out of this, a selection of product properties can be derived for the final product profile which becomes the basis for the definition of *the target specification.*

4.5.1   Implement Management System

4.5.1.1   Start with Selected Ecodesign Elements and Integrate Them into Existing Management Systems

For many enterprises, it may have initially seemed to be difficult and expensive work to install a management system. This is particularly so in a SME where a management system may not seem necessary at all. For smaller companies, a somewhat simpler approach could be to select a set of improvement rules and tools and apply them within an existing product development scheme. Those companies who already have a working management system in place can integrate an Ecodesign procedure therein.

In the end, the installation of a management system can have many advantages. One clear advantage is that, for many customers, the application of a management system according to ISO 9001 and ISO 14001 is already a precondition for supply. Other systems – already in the international standardization process – will be required soon, including energy management, EH&S management, and social responsibility.

Ecodesign fits best with or in the ISO 9001 standard, but can also be included in ISO 14001. In Annex 9, possibilities for integration are shown [cf. Sections 2.5 and 3.5].

> **Alert**
>
> A debate often arises between environmental and quality engineers about which management system (ISO 9001 or ISO 14001) is right for the integration of Ecodesign. This debate should be avoided, especially if these systems are already harmonized in the company within one integrated management system.

Having only one expert responsible for the whole Ecodesign process should also be avoided. This is a task for many experts. But after Ecodesign has been integrated into the development and production processes, the role of a coordinator might be required.

The necessary inputs for Ecodesign management in the E&E industry, aside from the structure of the Ecodesign directive, are described in IEC 62430 and, more generally, in ISO TR 14062. The checklist for integration of Ecodesign elements from IEC 62430 is included in Annex 10. The E&E industry can serve as a good example for other industries because they are the forerunners with the ERP. From the authors' point of view, the focus of the checklist in IEC 62430 is a little bit too close to the product. Therefore, we have added additional hints in the last two rows of the table in Annex 10. These hints widen the view from only the product to the performance of the product in the application system. Of importance is the total environmental impact of all comparable products sold worldwide! For SMEs, it might seem impossible to estimate such data, or a bit too theoretical. Furthermore, they may well believe they have very little, if any, influence. But internet searches and manufacturing organizations can often help to find sufficient information. The value of the information is obvious in those circumstances when arguments are needed to defend the product in any kind of political discussion. Also, an SME with an exemplary "green" product can change the whole market.

4.5.1.2 If No Management System Exists, Install Some Rules

In the beginning of the installation process for an Ecodesign management system, it is nearly impossible to do everything required in IEC 62430, or to follow completely all of the recommended rules. This complaint is heard not only from SMEs but also from the big companies. Philips started its "Ecovision" program with some simple design rules from the Ecodesign checklist named "*Quick 5*", the most important areas for reduction or improvement: weight, hazardous substances, energy, recycling and disposal, and packaging. Siemens currently uses *20 rules*. *However*, not all remain important when it comes to the final product design process. Without any further knowledge, a design engineer, without experience in applying such simple rules, can also achieve very good results.

The value underlying the completeness and quality of such simple checklists is experience! E&E industry experience has shown that between 50% and 90% of the impact of a product, over the whole lifecycle, is caused by energy consumption in the use stage. The application of nearly all simple improvement or reduction rules can lead to impact reduction potential which is not significantly different or better than that derived from more detailed and complex LCA investigations or scientifically complicated calculations. This dilemma is described in Fig. 4.16. An engineer can directly apply the rules to the product and its components.

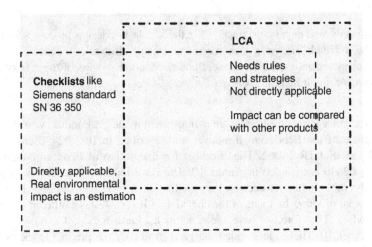

Fig. 4.16 Comparison of the improvement potential from the application of simple design rules and the improvement potential by a complete LCA to the Ecodesign of a new product (schematic example)

The LCA results, in comparison, usually only provide a summary of one set of values or impact categories, and must be carefully evaluated. Afterwards, it must be allocated, step-by-step, to the right component to identify the reduction potential. The advantages for the LCA tool are in its application to systems like plants or factories. As well, the data are more exact and detailed, and they offer additional improvement opportunities after the simple rules from the Ecodesign checklists have been applied.

In Fig. 4.16 an attempt is also made to show that no tool really covers all impacts. For the LCA of a product, the systematic problem lies in (a) the distribution of impacts to the functional units of the product, (b) whether the impacts are local, regional or global, and (c) more or less unknown processes in the life cycle, such as recycling. Another issue with LCA is that many of the 20–30 noted impact categories in ISO 14040 have not been scientifically accepted. So, the results can be misleading because they are often economically or politically motivated.

The above-mentioned simple rules are obviously generalized and could be applied to every product in the world. While the application of rules for simplification of disassembly can change the product structure, such changes are seldom the result of an LCA. On the other hand, simple rules cannot cover all of the total environmental impacts.

As experience with Ecodesign grows, companies will also investigate the LCA data of their components or products and use only those impact categories which are scientifically sound, such as the impact categories "global warming" and "ozone layer depletion". LCA data are also better for documentation and disclosures to the public. It should be mentioned that it has been agreed by the E&E industry associations and politicians in Europe that LCA data disclosed to the public should be independently validated. The consultants will, for example, check whether the model selected is acceptable and the data used are correct and not outdated.

So, as a minimum requirement for systematic application of Ecodesign, *Take the development process as it is, include some important environmental improvement design rules*, and if possible some simplified tools and start the new development process.

Alert

After some experience, the design engineer can recommend simple design rules *and* investigate complete LCA data to exhaust the total potential for improvement. [The discussion about non-misleading LCA data is complicated and will not be dealt with here. To avoid problems with some impact categories, the universal physical values in a Life Cycle Inventory can also be taken according to ISO 14041].

In the following, we describe a simple system that Siemens applies and from which a beginner can also easily take elements. After a certain amount of practice, most of the contents can, more or less, likely be applied by every E&E company. For Siemens, the elements are already integrated into the existing management systems.

Example

A simple System Based on Simple Rules [11]:

In the following example (till end of page 153), the Siemens system, which is a combination of simple design rules and detailed investigations, is presented.

The design rules cover the whole life cycle of a component or a product and can be easily added to the life cycle stages of the product. Here the responsible person can see who has to be involved in the task. In Fig. 4.17, as an example, some of these design rules are integrated in the life cycle.

In the first rule, a very rough estimation of the environmental impact over the whole life cycle is done. This is done by a rough calculation for some important phases as use phase. So it is avoided that the application of only experience rules doesn't bring enough improvement.

It must be mentioned that not all design rules have the same level of importance. In addition, every engineer is free to decide which rules are not relevant for his product. This freedom is important if only a component is being evaluated. However, every decision must be documented to ensure that decisions are explained. A similar approach can be taken for plants or projects, but obviously different design rules will have to be applied (cf. Annex 2b).

Marketing, planning, development
- estimate impact over life cycle, derive development targets
- integrate expectations of customers
- ....

Procurement, manufacturing
- reduce material consumption
- reduce weight
- ...

Sales and service
- information about disposal
- documentation for customers
- ...

Use/application
- information of long useful life
- ....

Disassembly, disposal
- ease of disassembly
- ....

Fig. 4.17 Simple design rules integrated in the life cycle (examples taken from [11])

The first question an engineer asks is:

What Should Be Done?

The answer is given in the first part of the standard SN 36 350-1 with its normative principles or rules. After some years of experience, it can be seen that every engineer asks similar questions. Where and how, for example, should I save energy? In which components do I find the hazardous substances?

On the other hand, the problem of how to distribute preferred solutions can arise? Where should all of the experience gathered after several years be placed, like the application of capacitors instead of batteries? A solution was the internal publication of a solution guide in which every design rule together with at least one example was presented to ensure common understanding:

How It Can Be Done?

An internet platform is very useful in extending the internal knowledge of the company and the engineers and good design solutions. A special team of experts from different branches of the company can serve as the evaluation and distribution team for the know-how behind the design rules and to update the Ecodesign standards.

It seems simple to say save energy! But in the detail, the many tricks or innovations should be described and should be stored in a brochure or on the company´s intranet for engineers not yet skilled with Ecodesign. So behind every design rule, a lot of know-how can be collected and, after some years, it becomes part of the company's intellectual property.

If a newly designed product is already developed, then another question arises:

What Has Been Achieved?

Is the new product better than the forerunner or the competitor's product? Does the improvement qualify as "best practice" in the company or in the world? Siemens developed an evaluation checklist similar to the self assessment methodology of business excellence used by The European Foundation for Quality Management [12] where, after the application of the corresponding design rules, the results are evaluated and assigned a score of between 0% and 100%. In this procedure, it was transferred to a scale of 0–100 points. The evaluation steps are described in Fig. 4.18.

With some additional explanation, an evaluation of the status of the implementation of all of the design rules may be possible. The sum of the points shows the relative environmental impact of the product. Rules that don't fit a specific component or product are left out, and the system is corrected and related to the remaining design rules. The evaluation can also be applied to the life cycle stages to show any environmental deficits. An example is shown in Fig. 4.19.

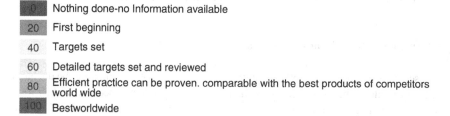

0	Nothing done-no Information available
20	First beginning
40	Targets set
60	Detailed targets set and reviewed
80	Efficient practice can be proven. comparable with the best products of competitors world wide
100	Bestworldwide

Fig. 4.18 Evaluation scheme for the assessment of the environmental performance of a product

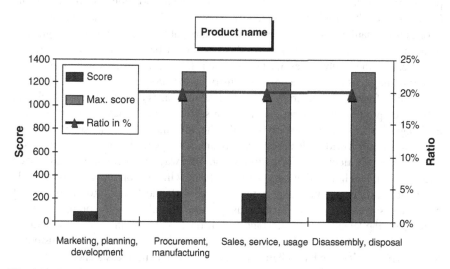

Fig. 4.19 Results of a self assessment of the Ecodesign of a new product evaluated over the life cycle stages

A self-assessment team consists of development engineers and all relevant experts. A corporate expert to calibrate the answers can also be useful. As the highest score possible or "the best solution in class" is the target, a continuous improvement of the product is possible with the same evaluation system. Some engineers might have problems with the "best in class" product. If this is really unknown, a comparison can be made with the forerunner product (previous model) or with the best competing product known.

As the engineer needs all information for Ecodesign, it is necessary to place all required information into the same toolbox. Within Siemens, this toolbox consists of the different parts of the standard SN 36 350 and additional checklists or guides. As the toolbox is open for further extension, further parts can be added. In the age of electronic information, all of the information is available on an internal homepage on the intranet. With the available "good practice" examples, questionnaires from customers can be answered and suppliers can be informed about any requirements. Parts of the information can also be directly placed on the external company homepage, such as the publicly available information about the way the company is conducting its Ecodesign activities and processes. One part of the toolbox for the engineer is the updated information about hazardous substances.

Information About Hazardous Substances

The second standard SN 36350-2 contains a normative list of substances to be declared or to be avoided. The substances published in this list are selected because of their relative hazard to health or the environment. Criteria are set in accordance with IEC Guide 113: Carcinogenic, Mutagenic, Toxic to Reproduction (CMR substances); chronically or acutely toxic; easily forming CMR or toxic substances; radioactive and water-polluting; persistent and bio-accumulative; contributing to global warming; ozone layer depleting. They will be accepted by suppliers as they are internationally agreed upon and standardized. Only potential and already published restrictions by a government are included. All substances on this list should be avoided. If this is not yet possible, then these substances must be declared. The lists are usually in harmony with corresponding lists of manufacturing (industry) associations.

For all information (materials in a product, substances to be avoided or declared, substances prohibited in some countries), a software tool should be used for data exchanges. It would be best to include the information in the order process. Unfortunately, a world-wide accepted software is not yet available. However, a system on how to declare information is on the way to becoming standardized (IEC 62474). It is hoped that large software firms will develop and offer worldwide compatible products. Integrated in the same tool should be an informative list of prohibited substances in the E&E industry. Although all have to comply with the same laws, the legislation on substances is difficult and contains many exemptions. The supplier, therefore, should also be informed about the requirements of his customer. Since SMEs don't often have chemical experts, they may need assistance. As the EU has the most extensive regulations on hazardous substances, the requirements flow therefrom. For a company that complies with the strictest regulations worldwide, it will be easier to comply with forthcoming regulations in other countries.

It is also helpful to have information on the components in which target substances are likely to occur. This additional information will be of special value to non-experts from purchasing departments when they prepare to buy components or materials. Such a list can also be the basis of information for customers, the public, for ordering, for development engineers, and for the application of any trend analysis to avoid risks.

It is often not known that a lot of restrictions have to be observed for packaging, and that packing materials are often seen as a part of the product. So another Siemens standard addresses environmentally compatible packaging:

Packaging Standard

In a packaging standard, the legal requirements and the recommendations of the company must be identified. Specific requirements could include: reusability of packing material as well as the complete package, made with renewable resources, avoidance of some materials and, occasionally, avoidance of any packaging what-soever! Packaging can at times constitute 5–10% of the price of a product. Also, fees for governmental packaging take-back systems can be avoided if there is no packaging applied at all. For example, packaging for some household appliances can be limited to a wooden palette and only a strip of plastic to fix the product!

Other Internal Standards in the Siemens Toolbox

A framework for the *material record* of a product has been standardized. In a special internal standard, the material declaration procedure was described and has been published as IEC PAS 61906. This avoided internal misunderstandings and with suppliers.

The *environmental product declaration* was also internally standardized, even though international standards were in existence. The reason was that, for Siemens, the self-declaration was important. Taking this as a framework, every member company of the Siemens Group is able to use the same standardized format to transmit comparable information to the public at the same level for all products declared.

How Is the Standard and the Toolbox Communicated Internally?

Besides the internal and external information on the internet, there is regular Promotion through seminars on internal procedures which are held for all interested experts and design engineers. An *internal environmental award* is also offered to get more good practice examples for development engineers in other groups.

4.5.1.3   The Product Profile Definition

The potential product profile can be derived according to the steps in Annex 10 after assessment of market conditions, related regulatory requirements, customer expectations, and comparisons with competitors. Another approach is to take a forerunner product (previous model) and examine missing or insufficient environmental properties and opportunities for environmental improvement. Additionally, an analysis of requirements over the product's life cycle (from the perspective of the supply chain, the application phase, reuse, and recycling) will result in further input to a potential profile. A planned reuse of components over several product generations requires very detailed investigations of costs and quality requirements.

The development engineers have to distinguish between "must" requirements, like the legal ones, and those already offered by competitors. From the latter, the requirements that are selected are those which fit into the overall framework concept for the new product development project. This is often guided by predefined cost limits. Once all of the aspects or requirements are set, the targets can be specified.

4.5.1.4   Implementation of Further Systematic Tools

Many tools have been developed in quality science in order to make related approaches systematic. As can be seen in the development scheme in Annex 10, systematic tools are not available in IEC 62430 for several of the steps. In this book, we offer some proposals on useful tools. They are listed in the scheme in Annex 10 in the last two rows. Those that are known from quality science, like QFD, are not described here. Also, the European Commission has started to offer Ecodesign tools.

The more information that is available, the more detailed investigations can be taken and further improvements initiated. In many cases, the following are design strategy alternatives (cf. Section 3.3):

1. *Reduce the complexity of a product by combining materials together* with types and by reducing the number of parts [Ideal target: only one type of insulating material (plastic), only one conducting material (copper)]. The framework condition is that the manufacturer will be willing to develop his own components. (cf. Section 5.2, part: simplification of the rack).

Plastics with a higher environmental impact can be checked to see whether they could be candidates for substitution by lower impact materials like polypropylene and then integrated into one component. The same can happen with different metal parts. The optimization is not done at the material level but rather at the component or part level. Therefore, it is possible that a component is produced with more expensive and less environmentally compatible material but, overall, the production process is cheaper, the volume of material required lower, and the component as a whole is environmentally better, and at a lower cost. Using this method of optimization, the focus is not necessarily on ease of disassembly nor reuse of components, but it might produce a longer lifespan.

2. Virtually dissect the product into the *essential functional components/units*[1] which can be individually optimized by LCA. Development of non-available components could happen together with suppliers. These components could also be especially designed for reuse [13].

Dissect the product virtually into only *commercially*[2] available components and optimize the product by using – if there is a possible choice between environmentally high and low impact components in the market – systematically low impact components to reduce the impact of the total product. This is a chance for a producer who assembles, but doesn't produce components. But this producer can also take the reduction opportunity that the market offers.

Environmental impact can be reduced by selecting components or by optimizing the software, e.g. loading behaviour, standby, etc.

Each of the strategic design approaches needs a systematic procedure to achieve the strategic target of minimizing environmental impact. CAD tools are usually applied by design engineers. Some tools are able to electronically use the environmental information on the components. The design software should also be able to assist the designer with the application of the different afore-mentioned strategies.

Also helpful are programs to optimize the assembly/disassembly times. Software for the automatic transfer of the material record of components has not yet been standardized. Therefore, some information will have to be gathered individually. In connection with the design strategies, a tool to *analyze reuse and recovery in advance* might be valuable. This kind of tool should start with an analysis of the *reusability of components*, or the whole product, and the potential value-added over the lifecycle. For quality requirements in this case, see IEC 62309. Applications of the components can also be for repair, as spare parts kept in stock, or for selling on the spot market. After the decision about planned reuse, an optimized design of the component or the complete product is possible.

This approach can also be applied in similar way to other products, such as motors. It contributes to the value added.

Material recovery can also be optimized for the same and other purposes. Within a strategy, it could be important to use valuable plastics instead of cheap plastics. In this case, it may be important to recover as much of the product as possible in

---

[1] In this case, the complete product is formed by functional units like filter, motor, heating, etc., and these units are combined and designed as easy-to-assemble/disassemble, often without further housings. The component can be designed, for example, for reuse if it contains only a part that doesn't wear or it contains parts of the housing. In [13], a vacuum cleaner is assembled by only functional units.

[2] In this situation there are, for example, producers of PCs. They design the housing and purchase everything else. In order to reuse components, the manufacturer of the PC has more chances because the used component is available in a broader market. The supplier of the component is responsible for the information about lifetime or availability in the market.

order to make recycling cost effective. The situation is similar for other valuable materials. For other tools, see the development scheme in Annex 10.

Based on the above-mentioned examples, it can be stated that a design engineer can achieve considerable Ecodesign success by only applying some simple design rules and a simple development scheme! Complexity can be taken out of the development scheme and the system can be kept reasonably simple. Further tools are helpful and can be applied with growing experience.

4.5.2 Extension of the Management System

As already mentioned, further management systems are being developed by standardization organizations. Standards on Health & Safety (OSHA 18000), and Social Responsibility (ISO 26 000) will both have influence on environmental properties or will be connected with environmental requirements and, in the IEC, work on Energy Management standardization has been started. Other management systems (e.g. chemicals management – see Sections 2.5 and 3.5 and innovation management – see Section 4.5.3) might have overlaps.

Nearly every manager in every plant would prefer to have and use a single and integrated management system. Companies that have already installed a so-called "reference process house" can harmonize their management system somewhat easier. They integrate sub-processes like Ecodesign in the "Lifecycle Management" process. Processes which do not fit the normal "reference process house" can be added as support processes, for example, to purchasing, to quality, or to environmental protection. In the following overview, the processes of a company are described at the highest level (Fig. 4.20).

Such a system can be structured into eight or more detailed levels. It is important for the integration of Ecodesign that all corresponding management activities are directly integrated into the operational or running process and not into the support process where other environmental activities are included. The flexibility of the "reference process house" model is that it can absorb all management systems. Parts of it can be additionally certified if necessary.

4.5.3 Target Specification Based on Product Profile

Targets should address the most important environmental impacts. Such target setting can be derived from the environmental product profile taken from EuP (cf. Section 1.2) where there is an exemplary description. The profile was discussed in some detail in Section 4.5.1.1.

The actual targets should be compared with those of competitors by benchmarking. Requirements to the height of targets can also be derived, for example, by the legislated climate protection programs (20–50% reduction). Further chances for reduction are combined with new innovative solutions, or the company can decide

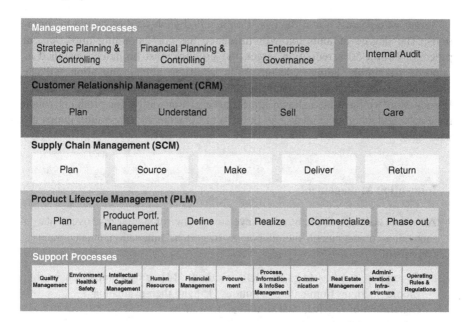

Fig. 4.20 Example of the Siemens Reference Process House

Aspects to be improved	Actual value	Target value	Result	Cost reduction
Energy consumption	....	....	− 7 %	....
Water consumption			− 15 %	
Chemicals consumption			− 40 %	
Noise			− 2 dB	
Total weight			− 21 %	
Number of parts			− 52 %	
Complexity			− 50 %	
Metals (weight)			− 22 %	
Plastics (weight)			+ 45 %	
Others (weight)			−55 %	
Packaging (types)			− 2	- 4 Million €
Packaging (weight)			− 84 %	- 0,1 Million €
Application of recyclates			50 %	(through reduced
Recycling degree/ ... Disassembly time			75 %/18 min	transport)

Fig. 4.21 Example for a Target Setting Profile [Schematic example based on an existing product. From the results it can be understood that high cost savings were possible. Cost savings are shown with only two exemplary examples]

to begin a factor X reduction program. Customers will "feel" the effect only if it saves them money. Therefore, the targets must be aggressive. The more ambitious the targets, the more the engineers will have to change the existing product.

Once defined, the targets have to be accepted by management. A simple profile should be developed to allow for ready comprehension and to make targets clear. In the following example, cost reductions can be estimated. Using such a simple scheme (as shown in Fig. 4.21), a decision by management can easily be taken.

Target values must not all be based on detailed analysis, but should be ambitious. The degree of ambition will influence the corresponding improvement process! With a 50% reduction target, the product must be newly designed as opposed to simply being redesigned. In addition, some innovation will be absolutely necessary (cf. the patent in Section 5.2, substitution of a fan). In the Introduction we have also shown the S-curve of innovation for products which have achieved a high technical age. In these circumstances, innovation jumps are necessary.

Innovation Management

As a necessary requirement to achieve strong environmental improvements, innovation management should not be forgotten. Innovation management consists of

- Generation of ideas (e.g. by WOIS method) [14] or by comparison with Bionic (learning from nature) solutions [15] and their evaluation
- Generation of practical test solutions from ideas
- Selection of the best alternatives
- Development of new technology
- Success analysis

It should be mentioned that some targets, like energy reduction in buildings, can also be promoted by government. Limits for stand-by and energy consumption of other E&E equipment will be standardized as part of the legal Ecodesign directive framework. Limits required by the Ecodesign directive have to be part of the development documentation.

4.5.4   Review of the Results

Besides an analysis of the results achieved, how the trends of environmental targets have changed and what has yet to be planned for the future should be reviewed. For reuse, a planning forecast is required for three product generations! But in order to plan for only one new generation, it might be wise to begin by installing of new technologies and production equipment.

Nobody really knows how ambitious target-setting can be for the next planning cycle if, for the last product iteration, an environmental improvement of, say 30% on average was achieved. Nonetheless, for the next generation, similar targets should be set. It is worth it for the environment.

Checklist for Management Action Plan – Section 4.5

Assessment questions	Answer	Comments	Recommended follow-up activities	Related sections
Is a decision taken for simple checklists and/or for a LCA approach?	Yes ☺ No ☹	Starting is easier with simple design rules	Find out where LCA has advantages; install LCA tool	
Is a toolbox available for a recommendation about hazardous substances, packaging, material records, environmental product declaration, recycling?	Yes ☺ No ☹	Engineers and experts require assistance	Extend tools if required	1.2, 2.5, 3.5
Are best practice examples available?	Yes ☺ No ☹	Experience must be available and collected in the enterprise	Install a systematic collection of good examples	3.3, 4.3
Are ambitious development targets set?	Yes ☺ No ☹	Awareness by customers is only achieved after strong improvements are made	Updates	2.5, 3.5
Is there a systematic approach to innovation?	Yes ☺ No ☹	To achieve very ambitious targets, new technologies and ideas are required. Patent situation can be improved	Check different innovation possibilities: Bionics, idea creation techniques	3.3, 4.3
Is there an evaluation system in place to measure progress of the design?	Yes ☺ No ☹	Evaluate results either by estimation or by LCA	Extend your practice	3.5
Are there rewards for the process of Ecodesign inside the company?	Yes ☺ No ☹	Without incentives, such as awards, the spread of ideas may be too slow	Change rewarding criteria to set a special focus to some criteria	4.1
Is your management system harmonized to integrate all required systems?	Yes ☺ No ☹	Many more management systems are in development	Avoid unnecessary systems, e.g., by discussion with customers	2.5, 3.5
Is a review process in place to continuously improve your Ecodesign?	Yes ☺ No ☹	Participation of CEO is required	To be reported annually	2.5, 3.5

Chapter 5
Examples

The purpose of this chapter is to provide useful examples of how one could address important issues such as CE marking and calculating the carbon footprint of products.

Section 5.1 describes conformity assessment requirements, including the CE marking requirement of the Ecodesign directive. An example which demonstrates an Energy Related Product – a complex set top box – has been chosen.

Section 5.2 provides an example of the successful implementation of Ecodesign from industrial practice.

Section 5.3 deals with environmental communication, by demonstrating how to calculate the carbon footprint of a product. The same product addressed in Section 5.1 is used to calculate the carbon footprint.

5.1 Conformity Assessment and CE Marking

Conformity means meeting the requirements specified in law and/or standards. Conformity assessment is any activity concerned with directly or indirectly determining that relevant requirements are fulfilled. Typical examples of conformity assessment activities are sampling, testing, and inspection; evaluation, verification, and assurance of conformity (supplier's declaration, certification); registration, accreditation, and approval as well as their combinations) [12].

CE stands for "Conformité Européene", a French term which means European Conformity. When the CE mark (see Fig. 5.1) is affixed to a product, it indicates that a manufacturer and the product is in compliance with the requirements stipulated in the EU machinery directive [1]. Since the Ecodesign directive was first introduced, environmental issues have been added for products affected by the Ecodesign directive. It is the sole responsibility of manufacturers that any CE

W. Wimmer et al., *ECODESIGN – The Competitive Advantage*,
Alliance for Global Sustainability Bookseries 18,
DOI 10.1007/978-90-481-9127-7_5, © Springer Science+Business Media B.V. 2010

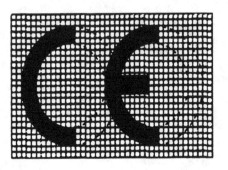

marking on their products meet the legal requirements and they bear ultimate responsibility for the conformity of the product. Environmental properties required can be declared by self-declaration. For technical purposes, self-declaration might not always be enough, i.e. for medical equipment. Then a certificate of a third party might be required.

In this context, conformity assessment relates to the design and production phases of the product. There are eight different conformity assessment procedures or "modules" which cover the design and production phases: internal production control (module A); CE type-examination (module B); conformity to type (module C); production quality assurance (module D); product quality assurance (module E); product verification (module F); unit verification (module G); full quality assurance (module H). The module applied to most ERP is module A, internal production control which covers internal design and production control.

One of the major requirements for manufacturers in fulfilling the CE marking requirements is to draw up a technical file (technical documentation). This document is intended to provide information on the design, manufacture, and operation of the product [2].

In the Ecodesign directive and accompanying implementing measures for a specific product category, it is stipulated that conformity assessment must be carried out and CE marking must be affixed to the product. This implies that the product under the conformity assessment requirement must comply with not only the requirements in the Ecodesign directive, but also in any other relevant directives, such as those related to low voltage devices (LVD) and electromagnetic compatibility (EMC).

5.1.1 Regulations and Studies of the EU About Set Top Boxes (STB)

Investigations on set-top boxes (STB) have led to a regulation on a simple STB [3]. Corresponding studies are available from the EU. Studies are also available for complex STB and regulation is planned [4]. It should also be mentioned that the requirements cover a conformity declaration according to Annexes IV or V of the Ecodesign directive. A copy of this text can be found in Annex 8.

Requirements, according to the Ecodesign directive for a manufacturer, are

– Documentation of the assessment result according to Annexes IV or V
– Conformity declaration according to Annex VI to be documented by
    (a) Results from Ecodesign activities
    (b) Comparison of achievements against requirements
– Documentation for recyclers
– Information for customers

Those companies having installed a management system according to Annex V of the Ecodesign directive or have been awarded an European ecolabel for their product are exempted.

To document the requirements for recycling, manufacturer organizations and the European Recycling Association offer standard checklists. Information to customers could be given by means of an environmental product declaration (in accordance with ISO 14025 or, more generally, ISO 14020).

In order to aid in understanding the conformity assessment requirements in the Ecodesign directive, the complex set top box (CSTB) was chosen as an example product for an ERP. A technical documentation file of the CSTB was developed in this section to demonstrate the applicability of the requirements to the CSTB.

5.1.2 Legal Requirements

In the Ecodesign directive, Article 5 stipulates the requirements for the CE marking and the declaration of conformity that an ERP must meet before its placement on the EU market. Of particular importance is the requirement on the declaration of conformity referred in an appropriate implementing measure. Specific elements of the declaration of conformity include according to the Ecodesign directive:

• The name and address of the manufacturer (or authorized representative)
• A description of the model
• The references for the harmonized standards applied, if appropriate
• Other technical standards and specifications used, if appropriate
• Reference to other Community legislation providing for the affixing of the CE mark that is applied, if appropriate
• Identification and signature of the person empowered to bind the manufacturer (authorized representative)

5.1.3 Procedures to Affix CE-marking

Since the implementing measures of the Ecodesign directive require module A for conformity assessment, the related requirements, together with the preparation of a technical documentation file, are discussed here.

Module A refers to internal production control where the manufacturer must [5]:

- Establish a technical documentation file, thereby enabling conformity of the product with the requirements of the directive to be assessed
- Keep the technical documentation file for at least 10 years after the last product has been manufactured
- Keep a copy of the declaration of conformity with the technical documentation
- Take all measures necessary in order that the manufacturing process ensures compliance of the manufactured products with the technical documentation

In general, the technical documentation file consists of two parts: essential technical data and supporting technical documentation.

(i) Essential technical data includes information such as authorized representative in the EU, manufacturer's name and address, manufacturer's declaration of conformity, certificates, product identification/description, list of harmonized standards/specification, and operating instructions.
(ii) Supporting technical documentation includes test plans, test reports, and engineering drawing, etc. [5].

There are two options for an assessment of conformity with the requirements of the applicable implementing measure: internal design control in Annex 4 or management system in Annex 5 of the Ecodesign directive. Both options are essentially the same. Differences between the two reside in which are the responsible party for managing and implementing Ecodesign, the design department or the environmental management department. The choice will be made by individual companies based on their own corporate structure and culture. In this book, internal design control was chosen for the conformity assessment. Key elements of the internal design control are summarized below [7].

A. A general description of the ERP and of its intended use
B. The results of relevant environmental assessment studies which are used by the manufacturer in evaluating, documenting, and determining product design solutions
C. The ecological profile, if required by the implementing measure
D. Elements of the product design specification relating to environmental design aspects of the product
E. A list of the appropriate standards
F. A copy of the information concerning the environmental design aspects of the product provided in accordance with the requirements specified in Annex 1, Part 2

G. The results of measurements on the ecodesign requirements carried out, including details of the conformity of these measurements as compared with the ecodesign requirements set out in the applicable implementing measure

The seven elements listed above comprise the specific requirements for the development of technical documentation file applicable to the ERP.

5.1.3.1 Developing Technical Documentation file of CSTB

A technical documentation file for the conformity assessment of a CSTB to the Ecodesign directive implementing measure is developed below. The conformity assessment procedure used in this example follows the internal design control in Annex 4 of the Ecodesign directive. For each of the seven elements, corresponding information must be filled out to complete the conformity assessment. The assessment results are inserted into the existing technical documentation file of the CSTB to meet the CE marking requirements of the CSTB.

A. A general description of the ERP and its intended use

1. Product definition [6].

   1.1. Definition of General CSTB
      A CSTB is a device that connects to a television and some external source of signal and turns the signal into content then displayed on the screen.

   1.2. Definition of target CSTB related to functions

      The target CSTB consists of the following functions

      • High Definition digital CSTB with internal mass storage media
      • DVB-S and DVB-S2 transmission standards (return path)

      1.2.1 High Definition digital CSTB with internal mass storage media

         High Definition television (HDTV) refers to the broadcasting of television signals with a higher resolution than the Standard Definition television (SDTB) with Hard Disk Drive (HDD)

      1.2.2 Type of Transmission platform (DVB-S receivers)

         Satellite CSTB: A CSTB whose principal function is to receive television signals from satellites and deliver them to a consumer display and/or recording device.

2. Scope definition

   Complex STBs are STBs which allow conditional access. A set-top box is a stand-alone device, using an integral or dedicated external power supply, for the reception of High Definition (HD) digital broadcasting services satellite and/or terrestrial transmission and their conversion to analogue RF and/or line signals and/or with a digital output signal.

CSTBs might have additional features, such as

– Return path/integrated modem/internet access
– Multiple tuners (for picture-in-picture or to serve several end-devices)
– Recording with internal mass storage media
– Entitlements

Digital receivers with a recoding function based on removable media in a standard library format (DVD, VHS tape, "Blue-ray" disc, etc.) are excluded from the scope, but complex STBs with removable media are included.

2.1 Target CSTB
   – Product name: Complex Set-top Box
   – Weight and volume: 3.712 kg including packaging and 380 × 60 × 300(mm)
   – Functionality: Receiving broadcasts, Personal recording and Multimedia, HD display
   – Company: Marusys (Korea)

Table 5.1 shows the part list and picture of each part.

3. Intended use
   3.1. Household

   Larger reception installations, such as communal satellite installations, which are not intended to serve one household only.

   3.2. Uses of product
      3.2.1. Transmission platform of media content delivery

         For satellite broadcasting, the signal is broadcast (often encrypted) by the media content provider to the satellite that transmits it back to Earth. The viewers need a satellite dish antenna to receive the signal and a CSTB to control the access to the signal (to decrypt the encrypted signal). The signal can be analogue or digital.

   3.3. Manual for consumer
   The manual for consumers include important information including:

   • Safety instructions – Important notes
   • Controls, displays, and connections
   • How to connect and set-up
   • Remote control information
   • First Installation
   • Operating instructions
   • On-Screen displays
   • EPG – Electronic Programme Guide

Table 5.1 Part list of CSTB

Part list			Quantity	Description
Set top box			1	Product
Packaging			1	Package and protect product
Main board			1	Control CSTB functions
SMPS			1	Control power supply
Cover top			1	Protect internal part
Panel assembly			1	Control function

(continued)

Table 5.1 (continued)

Part list		Quantity	Description
PCB panel assembly		1	Display information of CSTB
Cover back		1	Protect Internal part andventilation
Cable		1	Connection part

- Common interface
- Videotext (teletext)
- Edit channel list
- Channel search (channel scan)
- Parental control
- Settings
- Service menu
- Connecting up the video/PVR
- Using the front panel
- Technical appendix
- Short technical guide
- Troubleshooting

B. The results of relevant environmental assessment studies, which are used by the manufacturer in evaluating, documenting, and determining product design solutions

1. Results of typical power consumption

   The available market and product data – typical power consumption values can be estimated. The resulting typical power consumption levels in technical analysis of existing product (task 4) of the preparatory study are presented in Table 5.2.

2. Results of LCA on CSTB

   The total environmental impacts for CSTB according to the EcoReport calculations are listed in Table 5.3.

Table 5.2 Typical on mode and active standby power consumption of CSTBs per functionality

STB(DVB-C, -S, -T, IPTV)	Typical power consumption (W)	
	On mode	Active standby
Basic "complex" STB, including all typical interfaces, digit display, CI and/or CAM, etc.	10	5
Additional power consumption for	+8	+2
• Hard disk drive	+5	+1
• Second tuner/multiple tuners	+8	+8
• High definition capability	+10	+10
• Return path		

[6]

The use stage impacts on the total energy, electricity, and greenhouse gases for different modes:

– On mode: 55%
– Active standby mode: 45%
– Off mode: <0.1%

Materials acquisition, use, and end-of-life are all relevant. Regarding the on mode and active standby mode, both need further consideration, whereas the off mode is of minor relevance.

Typical power consumption of the existing CSTB gathered through a stakeholder questionnaire – Table 5.4 shows the survey results of the power consumption of CSTB [6].

C. The ecological profile, if required by the implementing measure.

"Ecological profile" means a description, in accordance with the implementing measure applicable to the ERP, of the inputs and outputs (such as materials, emissions, and waste) associated with an ERP throughout its life cycle which are significant from the point of view of its environmental impact and are expressed in physical quantities that can be measured (Ecodesign directive, Article 2 "definitions").

Table 5.5 shows a rough example of an ecological profile format of a product. The best approximation is the scheme from the base cases used for all ERP by the consultants.

There is no requirement to use Table 5.5 for all products but it is part of the investigation methodology published on the homepage of the EU [7]. A standardized methodology is not available. The European Commission is installing a revised LCA method where scientifically agreed impact categories are described.

Since the implementing measure of the CSTB does not require the declaration of the ecological profile of CSTB, there is no need to show the profile here.

In the event the manufacturer wishes to disclose the ecological profile, one can show a carbon profile as shown in Section 5.3 "Carbon footprint, carbon reduction opportunity and carbon management approach using set-top box as an example ERP". The carbon profile in Section 5.3 was developed based on

Table 5.3 Environmental impact of CSTB

Nr	Life cycle impact per product;							Date	Author		
0	Base Case 6: complex STB with HD, HDD, 2nd tuner, return path							0	KSchi		

Nr	Life cycle phases ->		Production			End-of-life			Disposal	Recycling	Total	Total
	Resources Use and Emissions		Material	Manufacturing	Total	Distribution	Use		Debit	Credit		Total
	Other Resources and Waste											
8	Total Energy (GER)	MJ	519	124	643	109	11,941		121	80	41	12,735
9	of which, electricity (in primary MJ)	MJ	154	30	184	0	11,936		0	35	−35	12,086
10	Water (process)	ltr	289	7	296	0	799		0	31	−31	1,063
11	Water (cooling)	ltr	131	34	165	0	31,827		0	7	−7	31,985
12	Waste, non-haz./landfill	g	17,778	268	18,046	79	14,018		1,806	100	1,206	33,349
13	Waste, hazardous/incinerated	g	538	3	541	2	280		712	39	673	1496
	Emissions (Air)											
14	Greenhouse Gases in GWP100	kg CO$_2$ eq.	31	8	39	9	521		9	5	4	573
15	Ozone Depletion, emissions	mg R-11 eq.	Negligible									
16	Acidification, emissions	g SO$_2$ eq.	349	41	389	27	3,077		18	29	−11	3,482
17	Volatile Organic Compounds (VOC)	g	2	2	3	1	5		0	0	0	9
18	Persistent Organic Pollutants (POP)	ng i-Teq	23	4	28	0	79		9	0	9	115
19	Heavy Metals	mg Ni eq.	216	10	226	4	207		34	4	30	467
	PAHs	mg Ni eq.	81	2	82	3	24		0	4	−4	107
20	Particulate Matter (PM, dust)	g	35	11	45	19	66		156	1	155	285
	Emissions (Water)											
21	Heavy Metals	mg Hg/20	222	0	222	0	79		10	20	−10	292
22	Eutrophication	g PO4	5	0	6	0	0		1	0	0	6
23	Persistent Organic Pollutants (POP)	ng i-Teq	Negligible									

Table 5.4 Typical power consumption data of existing CSTB

| STB (DVB-C, -S, -T, IPTV) | Typical power consumption (W) | |
	On mode	Active standby
Basic "complex" STB, including all typical interfaces, digit display, Cl and/or CAM, etc.	10	$4 \rightarrow 5$
Additional power consumption for	$+5 \rightarrow +8+5$	$+2$
• Hard disk drive	$+8$	$+1$
• Second tuner/multiple tuners	$+10$	$+8$
• High definition capability		$+6 \rightarrow +10$
• Return path		

[6]

Table 5.5 An example of an ecological profile format of an ERP (cf. Table 5.3)

Materials
Total (kg) of which
Disposal (kg)
Recycled (kg)
Other resources
Total energy (GJ)
of which, electric (in primary) (GJ)
Water (process) (m³)
Water (cooling) (m³)
Waste, non-hazardous/landfill (kg)
Emissions to air
GHG in GWP 100 (t CO_2)
Acidification Potential AP (kg SO_x)
Volatile Organic Compounds VOC (kg)
Persistent organic pollutants PoPs (mg i-Teq)
Heavy metals (mg Ni)
Polyaromatic hydrocarbons PAHs (mg)
Particular matter (dust) (kg)
Emissions to water
Heavy metals (g Hg/20)
Eutrophication (g PO_4)

the method related to the ongoing international standardization work on carbon footprinting [8].

D. Elements of the product design specification relating to environmental design aspects of the product

Energy consumption during the use stage was identified as the most significant environmental aspect; thus, this needs to be improved through product design. Product specifications based on the environmentally significant aspects were developed and are shown in Table 5.6.

Table 5.6 Product design specification related to the significant environmental aspects of CSTB

Life cycle stage	Type of specification		Name of specification	Explanation and value
Product use	Fixed	Power consumption not exceeding 15 W	Indication for energy consumption in the on mode	Typical power consumption (Basic CSTB, HD capability, Return path)
	Fixed	Power consumption not exceeding 9 W	Indication for energy consumption in the standby mode	Active standby mode

[6]

Table 5.7 Standards applied to the development of technical documentation file of CSTB

Standard name	Contents	Application area
EN/IEC 62301:2005	Measurement of electrical power consumption in the stand-by mode	Household Electrical Appliances
IEC 62087:2002/EN 62087:2003	Measurement of the power consumption of digital terrestrial, digital cable, and digital satellite CSTBs	Specific to digital television CSTBs with detailed converge of test signal and external loads
CEA-2013	Measurement and maximum limit of stand-by mode	Specific to digital television CSTBs includes treatment of parasitic peripherals such as security cards
IEC 62430	Environmentally Conscious Design of electro-technical products	

E. A list of the appropriate standards

Appropriate standards applied to the development of this technical documentation file are shown in Table 5.7.

F. A copy of the information concerning the environmental design aspects of the product provided in accordance with the requirements specified in Annex 1, Part 2. (Note: The requirements in Annex 1, Part 2 are requirements relating to the supply of information.) according to the Ecodesign directive.

Figure 5.2 shows the manufacturing process and picture of the CSTB.

The stand-by power consumption of <1 W printed on the receiver packaging is achieved only when the receiver is in stand-by mode with the display switched off. You can perform the setting in the following menu: "Main menu", "Settings", "Customize screen menu", "Front display in stand-by".

Do not switch the receiver off at the power switch directly from the operating mode. This can lead to loss of data and corruption of the software.

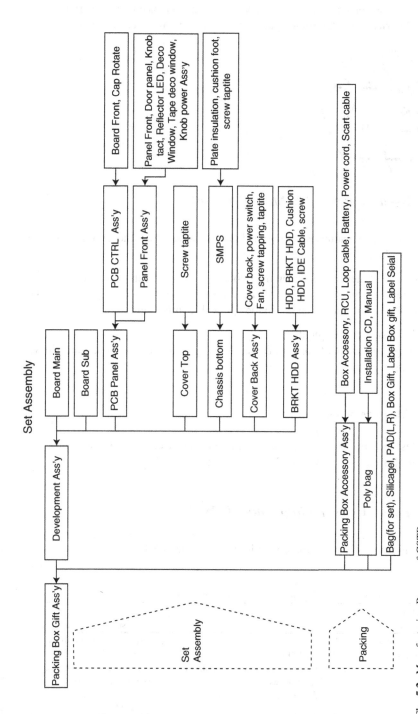

Fig. 5.2 Manufacturing Process of CSTB

Electronic equipment is not domestic waste – in accordance with WEEE directive on used electrical and electronic appliances, it must be disposed of properly.

5.1.4   Used Batteries Are Special Waste!

Do not throw spent batteries into your domestic waste; take them to a collection point for old batteries!

G. The results of measurements on the Ecodesign requirements carried out, including details of the conformity of these measurements as compared with the Ecodesign requirements set out in the applicable implementing measure.

G.1 Ecodesign requirements
Ecodesign requirements of CSTB were met by following the procedure shown in Table 5.8.

5.1.4.1   Measurements Method

Several test methods for the measurement of the energy consumption are shown in Table 5.9 of these methods, EN 62087 was used.

5.1.5   Power Consumption Data of CSTB

Power consumption data of the CSTB is shown in Table 5.10.

Table 5.8 Procedures for meeting the Ecodesign requirements of CSTB

D product design specification	E appropriate standards	G The results of measurements
Energy consumption during the use stage	IEC 62087:2002/ EN 62087:2003 Measurement of the power consumption of digital terrestrial, digital cable, and digital satellite CSTBs	Test results of the energy consumption during the use stage following EN 62087 certified by a laboratory accredited by KOLAS

[6]

Table 5.9 Measurement methods of the energy consumption of an ERP (n/a = not applicable)

	EN 62087	EN 62301	CEA2013-A[a]	CEA 2022
Scope	Specific to digital STBs with detail coverage of test signals and external loads	Not specific to STBs but detailed methodology on low power measurement	Specific to digital STBs	Specific to digital STB whose primary function is video reception and delivery
Measurement modes	Disconnected Off Stand-by passive Stand-by active (low) Stand-by active (high) On (play) On (record)	Stand-by Low Power mode	Sleep	On
Temperatures	15–35°C, with 20°C preferable	23 ± 5°C	22 ± 4°C	n/a
Power supply	Device rated voltage and frequency, ±2%	230 V ac/50 Hz/±1%	115 V RMS ± 3 V 60 ± 3 Hz	n/a
Instrument Accuracy	Not given	P≤10 W → 0.01 W 10 W < P ≤ 100 W → 0.1 W P > 100 W → 1 W	Resolution to be 0.1 W or better True power watt meter preferred	n/a

[a]CEA: The Consumer Electronics Association

Table 5.10 Power consumption of CSTB

Mode of operation	Power consumption (W)	Period (h)	Energy (Wh)
Watching digital broadcast			
- Free to Air broadcast (TV, Radio)	20.1	1.125	22.6
- Pay TV scrambled by conditional access system	21	1.125	23.6
			46.2
Personal Video Recorder			
- Single recording	21.1	1.125	23.7
- Dual recording	21	1.125	23.6
- Playing	20.8	1.125	23.4
- Trick play	21	1.125	23.6
			94.3
Multimedia			
- MP3	19.8	0.45	8.9
- JPEG	19.8	0.45	8.9
- GAME	19.8	0.45	8.9
- VLC streaming	19.8	0.45	8.9
			35.6
Channel search	18.9	0.45	8.5
Stand-by mode	9	14.76	132.8
Off mode	0	0.24	0
			141.3
	Total (per day)	24	317.6
	Total (5 years)	579.6 KWh	

5.2 Example – Siemens Former Mobile Phone Base Station

5.2.1 Introduction

The design of this particular product demonstrates that, not only are environmentally compatible solutions possible, but money can be saved when considering the whole life cycle. The product used in this example is the Siemens former mobile phone station model number 60/61 which is shown in Fig. 5.3. In Section 4.5, the Ecodesign system of Siemens is described in some detail, and serves as an example of a simple design procedure that may be applicable to every SME. Similar design rules as identified in Annex 2a were applied for this solution.

5.2.2 Target

The target of the design solution was to address the main environmental impacts: energy consumption, resources required, and radiation impact. While the radiation was not particularly high, it was reduced because of general public concern.

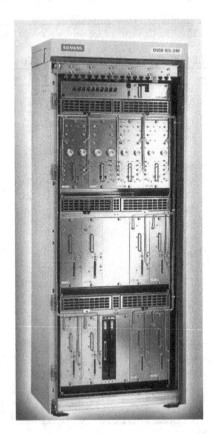

Fig. 5.3 Base Station BS 60/61

5.2.3 Examples for Improved Environmental Solutions

5.2.3.1 The Cabinet (Containing Electronics)

When considering the environmental improvement opportunities for a cabinet, there are two areas in which environmental optimization can be pursued: (i) the structure or design of the cabinet and (ii) the materials used.

In terms of structure, one can optimize the use of the *volume*. Such a volume optimization may not always appear to make sense for industrially applied products if resource costs do not play a role. However, in this case, a total volume reduction could be achieved and, within the same volume, eight transmission units could be placed instead of six and resources were therefore saved.

The selection of the housing material offers different optimization opportunities, depending on climatic conditions and different marking technologies. Stainless steel was selected for the housing in this case. While stainless steel is more expensive than other materials, it is inherently resistant to corrosion and fire, it does not change colour, and inscription can be done by a laser. The forerunner previous model consisted of chemically oxidized aluminium which was also coated. The marking was printed, which required additional chemicals. A total cost reduction of 50% was achieved in comparison with the previous model.

Fans are often required for cooling electronic components as elevated temperatures can reduce reliability. The right structure, or design, can be enough to create "natural" circulation of the heated air, thereby reducing or eliminating the need for a fan. The climatic conditions in the location where the base stations are to be located are often not known in advance, or are often not considered or investigated. Therefore, it was decided to have one "robust" model for a variety of applications. For an outdoor application of a mobile phone base station, air conditioning in up to 55°C outside temperatures is possible without the need to include active cooling. The energy consuming fan is thereby avoided.

5.2.3.2 Energy Consumption

As energy costs have dramatically increased over the last few years, energy consumption is increasingly of interest. For a telecom provider, the number of stations (many thousands) can be multiplied by the energy consumption of each piece of equipment to determine the total costs. Such an investigation is also necessary to determine the environmental impacts of such equipment as energy consumption and the emission of greenhouse gases can be easily correlated. As a result, energy consumption has become a political issue as well. For the BS 60/61, when compared to the former product generation, a 35% reduction of the power uptake could be achieved. For the first year of sales alone, the energy savings equate to a reduction of some 57,000 t of CO_2 and about €6 million of cost reduction for customers. This would be like conserving, or simply not using 503 train wagons of coal to generate electricity if the tons of CO_2 saved are converted to a standard load of a coal wagon.

5.2.3.3   Electronic and Software Oriented Tools

In the case of a mobile phone base station, the tele-recognition of the "Electronic serial number" can supply all of the configuration values and can enable the optimization of service intervals. The environmental benefits are reduced travel for service engineers and stability of operating characteristics (especially energy consumption which stays at the same level). In total, tele-recognition results in reduced energy consumption and lower use of auxiliaries and resources.

Reduced expenditures for administration and maintenance with implementation, documentation, service, software updates, and inventories by customers can be achieved by the software via the internet. Also, recycling of the plant can be planned or the technological or software state of the system can be checked: Is the energy consumption too high compared with a new product generation or could new software be implemented which saves more energy?

The role of software for energy consumption is frequently underestimated. Much more energy could be reduced if the software was optimized, e.g., by avoiding battery loading or stand-by. Additionally energy or resource saving components, like capacitors instead of batteries, could be used.

5.2.3.4   Improved Reception Sensitivity

The discussion about radiation protection will undoubtedly continue. Optimization is electronically possible. A 37% reduction of transmission impact of all cellular phones in the net can be achieved with such a base station. In the end, for the users of cellular phones, a reduced radiation impact is the result. As a further benefit, the application time of the phone is increased as the life span of the batteries of the cellular phones became longer.

5.2.3.5   Innovative Ideas Can Be Found Everywhere

As mentioned above, for outdoor applications no active cooling is necessary. The advanced cooling by membrane filter was the innovative approach that received an EU patent. The package density was increased for the cabinets and racks, also the cooling was improved. The following savings were achieved because of the new heat exchanger system:

- 7 K better heat balance
- MTBF – improvement 31%
- Cost reduction 33%
- Weight reduction 50%
- Volume minus 38%
- Energy consumption: –180 W

As the product optimization was often only promoted for cost reduction in manufacturing processes or the application of low-cost parts, high levels of innovation for energy and resource reduction are still possible.

5.2.3.6 Parts and Types Reduction

In every cost reduction programme, the reduction of parts and types is reviewed. From an environmental point of view, such a reduction process is critical. In a normal parts and types reduction programme, there is a distinction made between the 20% most applied components or products and the remaining 80% which should be reduced or omitted. But from the environmental point of view, the target could be to reduce the type of materials used to one kind only (e.g. one type of metal or plastic for all). This target can be approached by evaluating the materials to be substituted and, for example, replacing more expensive high impact plastic with low impact ones with better environmental performance. A further optimization could be to combine several parts into one part (e.g. in an injection molding process) with the result being fewer parts and ideally made from only one type of plastic or metal. Therefore, the target is much more ambitious than in normal types and parts reduction exercises. The resulting environmental impact and the costs can be much lower than before. Assembly and disassembly costs will be reduced and recycling costs will be replaced by earnings.

If we look at the old version of a sub-rack in the mobile phone base station (Fig. 5.4) where the printed wiring boards were inserted, we find 66 different parts, four different materials and 25% more space required than for the new solution (Fig. 5.5).

The key features of the new sub-rack are:

- Number of different parts is now 17 (4 sheet metals, 1 support part, 12 screws).
- Design complexity: four different parts.
- Integrated parts: printed wiring board, air conditioning lattice, fixation of backplane positioning.
- Environmental advantage: one common material for all. Cost: 78% less than the former sub-rack.

Fig. 5.4 Old solution (sub-rack)

Fig. 5.5 New solution (sub-rack)

Here a discussion between quality engineers and environmental engineers becomes interesting – how is quality determined? The "Quality" of the old solution may be 100% but this won't be 100% satisfactory if the complexity of the old solution is evaluated.

5.2.4 Summary: Results for Mobile Phone Base Station

Sensitivity: + 2 dB higher = Power consumption 37% less is possible in mobile phone
New cooling system: (33% less cost)

- Cooling by air without active cooling
- Cooling with membrane filter (= no heat exchanger)

Front: Noble steel (frottage structure), laser inscription: 100% recycling possible
Subrack:

- New Solution: one material, only 17 parts, ca. 80% lower cost
- Old solution: 66 parts, four different materials, 25% more space required

Service: Costs reduced by remote control
Packaging: Now plug & play from factory (= less packaging); only wood as packaging material and multiuse packaging
Total product: Nearly 100% recycling possible. High cost reduction for the overall product achieved

5.3   Carbon Footprint, Carbon Reduction Opportunity, and Carbon Management Approach Using Complex Set-Top Box as an Example ERP

In accordance with the procedures and methods described in Section 2.3.1.3, a carbon footprint of a product was developed, carbon reduction opportunities were identified, and a carbon management approach was discussed using a complex set-top box (CSTB) as an example ERP.

5.3.1 Developing Carbon Footprint

In order to develop the carbon footprint, the CSTB was disassembled into major parts and subparts and the weight of each part and subpart measured. The Bill of Material (BOM) data for the CSTB was used to aid in the disassembly. The information gathered in this step was used to formulate the product structure by identifying the various processes, ranging from the use of raw materials to the manufacturing stage of the CSTB. In addition, scenarios for the distribution, use, and end of life cycle stages were made based on realistic data and reasonable assumptions. This included data for the transport mode and distance during the distribution stage, time for the use of the product during the use stage, and the collection distance, mode of recycling and disposal of the product during the end of life stage. Combining both product structure and scenarios comprised the system boundary of the CSTB. This process is often termed "product modelling" and in this case the result provided a basis for the development of the carbon footprint and subsequent identification of the carbon reduction opportunities, as well as an approach for the carbon management of the product. The system boundary of the CSTB is shown in Fig. 5.6.

According to the system boundary, all of the materials used for the six main components were considered. No cut-off was applied since it was possible to identify all materials. Table 5.11 is the material composition of the CSTB based on the BOM. In addition, it includes the GHG emission factors of each material. The factors were calculated based on the method in the IPCC Guideline [9] by processing the LCI data of each material in the Ecoinvent v 2.1 [10] and other relevant databases.

There are several processes involved in the manufacturing of the parts and the product itself. They are manual insert, auto insert and surface mounting technology (SMT) for the production of printed wiring boards (PWB), and set assembly of the CSTB. Data for the PWB and development assay are site-specific data. In addition, there were inputs into the manufacturing processes, such as steam and hot water. However, their consumption was negligible compared with other input. Table 5.12 shows the energy consumption data in the manufacturing processes.

Data for the distribution stage includes transportation from the CSTB assembly plant located in Dongtan, Korea to a port in Busan by truck and then to Denmark by sea. Table 5.13 shows the scenario for the distribution stage of the CSTB and the GHG emission factors for the different transport modes.

For the use stage of the CSTB, the electricity consumption can be calculated assuming that the use pattern of the CSTB is the same as the standard use pattern defined by Version 2.0 of the ENERGY STAR Program's "Requirements for Set-Top Boxes" [11]. Based on this assumption, the daily use pattern of the CSTB is 9 h in on mode, 14.76 h in standby mode, and 0.24 h in off mode, and the lifetime of the CSTB is 5 years. Table 5.14 shows the energy consumption data during the use stage of the CSTB.

The scenario for the end-of-life stage is based on the nature of the material: recyclable, combustible, and inert (neither recyclable nor combustible). As shown in Table 5.15, the CSTB consists of 67.2% recyclable materials, 27.4% combustible materials, and 5.4% inert materials. These figures, however, are theoretical because not all CSTB discarded are collected. Thus, a 75% recovery rate of CSTB in the EU's WEEE directive was assumed to apply. As such, 25% of the CSTBs discarded are assumed to be unaccounted for. In other words, the whereabouts of the unrecovered 25% of CSTBs is unknown. Table 5.15 shows the end-of-life stage scenario with the GHG emission factors of each material.

Summing up all of the data over all the life cycle stages of the CSTB constitutes the product modelling result of the CSTB, and the result is shown in Table 5.16. Normally material, energy, process, transport, and activity data are termed inventory data.

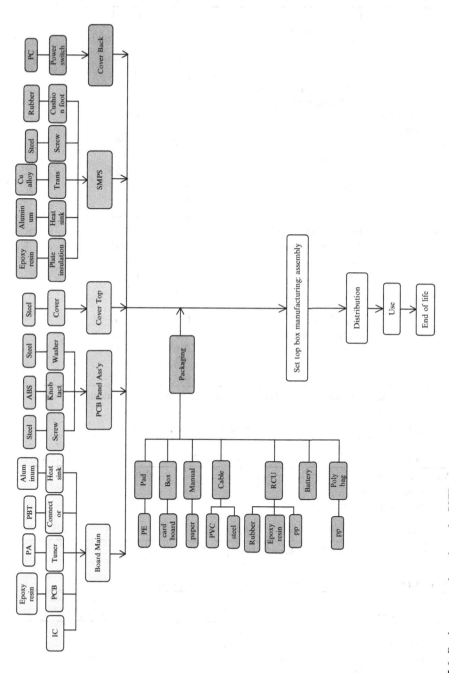

Fig. 5.6 Product system boundary of a CSTB

Table 5.11 Composition of the materials with the corresponding GHG emission factors of the CSTB

Component name	Material	Mass (kg)	GHG emission factor (kg CO_2-eq/kg material)[a]
Board main	Epoxy resin	0.406	6.7304
	PBT (Polybutylene terephthalate)	0.1	2.7011
	PA (polyamide 66)	0.052	8.0191
	Aluminium	0.014	12.376
	IC (integrated circuit component)	0.004	1012.4
PWB(printed wiring	ABS (Acrylonitrile butadiene styrene copolymer)	0.148	4.4031
board) Panel Assembly	Steel	0.0187	4.4768
Cover top	Steel	1.1247	4.4768
SMPs	Epoxy resin	0.284	12.376
	Cu alloy	0.07	1.8926
	Aluminium	0.048	12.376
	Rubber	0.02	2.6531
	Steel	0.007	4.4768
Cover back	PC (Polycarbonate)	0.01	7.7876
Packaging	Cardboard (corrugated)	0.632	1.1403
	Paper	0.24	1.6851
	PVC (Polyvinyl chloride)	0.17	1.9981
	Steel	0.182	4.4768
	Rubber	0.02	2.6531
	Epoxy resin	0.014	6.7304
	PP (polypropylene)	0.042	1.9825
	Battery (component)	0.016	101.84
	PE (polyethylene)	0.088	1.9485 (2.702)
	Silica gel	0.002	2.7091

[a]Derived from the LCI database, Ecoinvent v 2.1 following the IPCC 2007 calculation method

Table 5.12 Manufacturing stage energy consumption data of the CSTB

Manufacturing process	Electricity (kWh)	GHG Emission factor (kg CO_2-eq/kWh Electricity)[a]
Manual insert process	0.45	0.495
Auto insert processing	7.5	
SMT processing printed wiring board mounting facilities	8	
Set assembly	0.14	

[a]Derived from the Korean LCI database, following the IPCC 2007 calculation method

Table 5.13 Distribution stage scenario of the CSTB and the GHG emission factors

Transport mode	Distance (km)	GHG Emission factor (kg CO_2-eq/t-km)[a]
40 t truck (from plant to Busan) transport, lorry >32 t, EURO5	364	0.10402
Vessel (from Busan to Denmark) Transport, transoceanic freight ship	20,555	0.010749

[a]Derived from the LCI database, Ecoinvent v 2.1 following the IPCC 2007 calculation method

Table 5.14 Use stage data of the CSTB

	Power consumption (W)	Hour (h)	Energy (Wh)
On mode	20.28	9	182.52
Stand-by active mode	9	14.76	132.84
Off mode	0	0.12	0
	Total (per day)	24	315.36

Table 5.15 End-of-life stage data of the CSTB with the corresponding GHG emission factors

Mode of operation	Material	Weight (kg)	GHG Emission Factor (kg CO_2 – eq/kg)[a]
Incineration (27.4%)	Epoxy resin (disposal, plastic, consumer electronics, 15.3% water, to municipal incineration, CH)	0.704	3.029
	Rubber (disposal, rubber, unspecified, 0% water, to municipal incineration, CH)	0.04	3.1388
	PVC (disposal, polyvinylchloride, 0.2% water, to municipal incineration, CH)	0.17	2.2611
	PBT (disposal, Polybutylene terephthalate, 0.2% water, to municipal incineration, CH)	0.1	2.033
	IC (disposal, capacitors, 0% water, to hazardous waste incineration, CH)	0.004	2.5017
Recycle (67.2%)	Steel[+]	1.3324	−1.69
	Copper alloy (copper, secondary, from electronic and electric scrap recycling, at refinery, SE)	0.07	0.10399
	Al (aluminium, secondary, from new scrap, at plant, RER)	0.062	0.41924
	PC[+]	0.01	−0.332
	ABS[+]	0.148	−0.332
	Card board (corrugated board, recycling fibre, single wall, at plant, RER)	0.632	0.98533
	Paper (paper, recycling, with de-inking, at plant, RER)	0.24	1.5564
Landfill (5.4%)	Silica gel	0.002	0
	PE[+]	0.088	0.491
	PP[+]	0.042	0.415
	PA (Nylon 66)	0.052	0.0897
	Battery	0.016	0.797

[a]Derived from the LCI database, Ecoinvent v 2.1 and BUWAL250 (marked [+]) following the IPCC 2007 calculation method

Table 5.16 Product modelling result of the CSTB

Life cycle stage	Inventory data	Data source
Use of raw materials	Steel (1332.4 g), Epoxy resin (704 g), Cardboard (632 g), Paper (240 g), PVC (170 g), ABS (148 g), PBT (100 g), Copper alloy (70 g), Aluminium (62 g), PA (52 g), PC (10 g), PE (88 g), Rubber (40 g), PP (42 g), Silica gel (2 g), IC (4 g), Battery: AAA size(16 g)	Measured
Manufacture	Man Insert (0.45 kWh), Auto Insert (7.5 kWh), Surface Mounting Technology (8 kWh), Set assembly (0.14 kWh)	Measured
Distribution	Distance: 364 km by 40 t truck (from plant to port), 20 555 km by vessel (from Busan to Denmark)	Measured
Use	Energy consumption (5 years): 575.5 kWh	Scenario
End of life	Recycle: [Steel (1332.4 g), Cardboard (632 g), Paper (240 g), ABS (148 g), Copper alloy (70 g), Aluminium (62 g), PC (10 g)] × 0.75	Scenario
	Incineration: [Epoxy resin (704 g), PVC (170 g), PBT (100 g), Rubber (40 g), IC (4 g)] × 0.75	
	Landfill: [Silica gel (2 g), PE (88 g), PP (42 g), PA (52 g), Battery (16 g)])] × 0.75	

The next step is to calculate the GHG emissions from each life cycle stage shown in Table 5.16. This is determined by taking the inventory data in Table 5.16 and multiplying by the corresponding GHG emission factor. Fig. 5.7 shows the calculation procedure.

The GHG emission results of the CSTB calculated in Fig. 5.7 are shown in Table 5.17.

The CO_2 equivalent results in Table 5.17 have two applications. One is for the development of the carbon profile of the product and the other is for the identification of the carbon reduction opportunities of the product over its entire life cycle. As for the carbon profile, the Table 5.17 results can be envisaged as the carbon profile of the product. Depending on the definition of the carbon footprint, a single score which is the sum of all the CO_2 equivalent values in Table 5.17 can be considered as the carbon footprint of the product. In this case, the carbon footprint of the CSTB is 318.8 kg CO_2-eq.

The carbon reduction opportunity can be identified by applying the contribution analysis method to the carbon footprint results in Table 5.17. Every entry in each cell of the carbon footprint matrix is divided by the total sum of the CO_2 equivalent value (318.8 kg CO_2-eq) and expressed as the percentage value. The percentage value of each entry represents the relative contribution of the entry to the total CO_2 equivalent value of the product. Any contribution from an entry greater than x%, say 1%, can be used as a criterion for selecting the significant parameters requiring improvement. Applying this criterion to the results in Table 5.17, one can choose significant parameters which offer opportunities for carbon reduction. They include the use stage, manufacturing stage, and the steel used during the use of raw material stage of the product.

I. Calculation for the use of raw materials stage (extraction of resources from nature to production of raw and ancillary materials)

GHG emissions of the use of raw materials = $EF_{epoxy\ resin}$/kg x 0.704 kg/epoxy resin + EF_{PBT}/kg x 0.1 kg/PBT +EF_{PA}/kg x 0.052 kg/PA + EF_{steel}/kg x 1.3324 kg/steel + $EF_{Cu\ alloy}$/kg x 0.07 kg/Cu alloy + $EF_{aluminium}$/kg x 0.062 kg/aluminium + EF_{PC}/kg x 0.01 kg/PC + EF_{ABS}/kg x 0.148 kg/ABS + $EF_{cardboard}$/kg x 0.632 kg/cardboard + EF_{paper}/kg x 0.24 kg/paper + EF_{PVC}/kg x 0.17 kg/PVC + EF_{PE}/kg x 0.088 kg/PE + EF_{Rubber}/kg x 0.04 kg/Rubber + EF_{PP}/kg x 0.042 kg/PP + EF_{IC}/kg x 0.004 kg/IC + $EF_{Battery}$/kg x 0.016 kg/Battery + $EF_{Silicagel}$/kg x 0.002 kg/Silicagel

II. Calculation for the manufacturing stage
GHG emissions of the manufacture stage = $EF_{electricity}$/kWh x (0.45 kWh/ manual insert process + 7.5 kWh/ auto insert process + 8 kWh/ SMT process + 0.14 kWh/set assembly process)

III. Calculation for the distribution stage
GHG emissions of the distribution stage = EF 40 ton truck/(ton-km) x distance travelled (364 km) x set-top box weight(3.7124kg) x 1ton/1000kg + EF vessel/(ton-km) x distance travelled (20555km) x set-top box weight(3.7124kg) x 1ton/1000kg

IV. Calculation for the use stage
GHG emissions of the product use stage = $EF_{electricity}$/kWh x 0.31536 kWh/day x 1825 day

V. Calculation for the end of life stage
GHG emissions of the end-of-life stage = ($EF_{incineration,\ epoxy\ resin}$ / kg X 0.704 kg/epoxy resin + $EF_{incineration,\ PVC}$ / kg X 0.170 kg/PVC + $EF_{incineration,\ rubber}$ / kg X 0.04 kg/Rubber + $EF_{incineration,\ IC}$ / kg X 0.004 kg/IC + $EF_{incineration,\ PBT}$ / kg X 0.1 kg/PBT + $EF_{recycling,\ steel}$ / kg X 1.3324 kg /steel + $EF_{recycling,\ copper\ alloy}$ / kg X 0.07 kg/copper alloy + $EF_{recycling,\ Al}$ / kg + 0.062 kg/Al + $EF_{recycling,\ PC}$ / kg X 0.01 kg/ PC + $EF_{recycling,\ ABS}$ / kg + 0.148 kg/ ABS + $EF_{recycling,\ card\ board}$ / kg X 0.632 kg/card board + $EF_{recycling,\ paper}$ / kg + 0.24 kg/paper + $EF_{landfill,\ silicagel}$ / kg + 0.002 kg/Silicagel + $EF_{landfill,\ PA}$ / kg + 0.052 kg/PA + $EF_{landfill,\ PE}$ / kg X 0.088 kg/PE + $EF_{landfill,\ PP}$ / kg + 0.042 kg/PP + $EF_{landfill,\ battery}$ / kg X 0.016 kg/Battery) x 0.75

Fig. 5.7 The calculation procedure of the GHG emissions of the CSTB. Note: EF=emission factor

The carbon footprint in Table 5.17 can also be plotted with respect to the life cycle stages as shown in Fig. 5.8. The use stage was identified as the most significant life cycle stage followed by the use of raw material stage. Both life cycle stages offer carbon reduction opportunities.

Carbon management means the control of the GHG emissions of materials, parts, processes, and activities of a product over its entire supply and value chain. The carbon reduction opportunities identified above show which materials, parts, processes, and/or activities should be controlled and managed. The practical means

Table 5.17 The GHG emissions of the CSTB (kg CO_2-eq/fu)

Use of raw materials		Manufacture		Distribution		Use		End of life		
Materials	GHG	Manufacture	GHG	Delivery	GHG	Use	GHG	End of life		GHG
Steel	5.96E + 00	Manual insert	2.23E – 01	By 40 t truck	1.41 – 01	Electricity	2.85E + 02	Waste (incineration)	Epoxy resin	1.60E + 00
Epoxy resin	4.74E + 00	Auto insert	3.71E + 00	By vessel	8.20 – 01				Rubber	9.42 – 02
ABS	6.52 – 01	Surface mounting technology	3.25E + 00						PVC	2.88 – 01
PBT	2.70 – 01	Develop assembly	6.93 – 02						PBT	1.52 – 01
Copper alloy	1.32 – 01								IC	7.51 – 03
Aluminium	7.67 – 01							Recycle	Steel	-1.69E + 00
PA	4.17 – 01								Copper alloy	5.46 – 03
PC	7.79 – 02								Al	1.95 – 02
Card board	7.21 – 01								PC	-2.49 – 03
Paper	4.04 – 01								ABS	-3.69 – 02
PVC	3.40 – 01								Card board	4.67 – 01
PE	1.71 – 01								Paper	2.80 – 01
Rubber	1.06 – 01							Landfill	Silica gel	0.00E + 00
PP	8.33 – 02								PE	3.24 – 02
IC	4.05 – 00								PP	1.31 – 02
Battery	1.63E + 00								PA (Nylon 66)	3.50 – 03
Silicagel	5.42 – 03								Battery	9.56 – 03
Total	20.5296		11.2183		0.9608		284.8874			1.2443

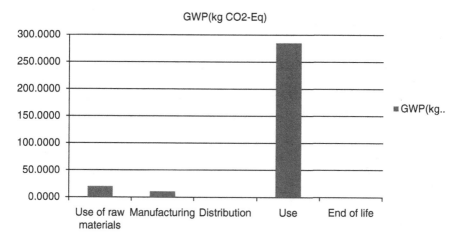

Fig. 5.8 Carbon footprint per life cycle stage of CSTB

for such control is to manage the materials and parts suppliers. Existing supply chain management systems can be utilized for this purpose. As to the control and management of the processes and activities, consumer education, together with appropriate governmental intervention such as carbon taxation or carbon labelling, could be considered.

Chapter 6
Outlook: Sustainability – What Does the Future Hold?

This book has attempted to help the reader prepare for the future – to come up with better products, to adapt the way we do business, and to incorporate a way of thinking that will facilitate tomorrow's success.

But what will the future bring? Certainly a question that is difficult to answer.

6.1 Upcoming Challenges

If we look at the current and projected trends laid out in the previous chapters, we can see some developments that are foreseeable. There are at least three main challenges companies will have to deal with:

- Climate issues: How to reduce greenhouse gas emissions?
- Energy issues: How to reduce energy consumption and how to become independent from fossil-based energy supply?
- Resource issues: How to manage resources, how to design and run smart resource cycles?

Some specific examples:

- Water-intensive business models will increasingly need to focus on minimizing their consumption patterns.
- We can no longer continue to use such a significant amount of fossil-based energy. Every step to increase energy independence is a worthy investment that will also help secure a company's business future.
- Finding ways to keep resources in circulation will be a key challenge to master (e.g. through establishing take-back systems for products to reuse parts and components).
- The carbon footprint of products will become more and more important. While we can already see some efforts in calculating and communicating the carbon

W. Wimmer et al., *ECODESIGN – The Competitive Advantage*,
Alliance for Global Sustainability Bookseries 18,
DOI 10.1007/978-90-481-9127-7_6, © Springer Science+Business Media B.V. 2010

footprint of products, we can easily foresee a time in the near future when this instrument becomes really important. It will most likely be used as a tool to not only improve products by reducing the carbon footprint but also through communicating it as a competitive advantage to more and more environmentally aware markets.

These challenges are already in sight but there will certainly be even more issues that will demand attention.

6.2 Consequences

From an understanding of the upcoming challenges, there are some consequences one can foresee. Preparing to deal with these consequences will take the form of guiding principles that will be used to help companies make informed decisions. The following consequences or principles can be seen:

- Low energy
  Implement energy efficiency in products, processes, and facilities.
- Low carbon
  Reduce the carbon footprint of products, processes, and facilities. Develop and use renewable energy sources.
- Low distance
  Decentralize the value chain of the product and reduce dependency on centralized suppliers or on extensive transportation.
- Low pollution and waste
  Avoid pollution and generation of waste – waste is an indicator of inefficiency.

6.3 Conclusions

Product sustainability and Ecodesign should be understood as an ongoing process, always setting new targets to achieve. The systematic approach described in this book should help integrate new thinking into management systems and prepare the company for more stringent environmental requirements over time. Reduction targets for products will become higher and new regulations, lack of resources or new energy crises can be expected. But the company, the business model and finally the products can be prepared for these changes. In this book we have tried to demonstrate that there may be different pathways to becoming an environmental champion. There are options for what might best fit the culture of each company. We have proposed tools to help companies apply more systematic approaches, taking into account upcoming standardization and regulations. However, when it comes to the details some things will change and become more precise over time, such as with product carbon footprint (PCF) or with the Ecodesign directive.

In the future, the Ecodesign directive requirements will regulate the environmental impacts of nearly all products. As a result, Ecodesign directive was one of the more important subjects dealt with. Complying with the Ecodesign directive criteria will also help prepare for the CE mark requirements. But one should not be limited to the existing Ecodesign directive requirements! In the future, these will become only the basic requirements.

The fundamental philosophy should be that, aside from all of the directives and regulations, innovation, creativity, and flexibility is required for product design. The ability to innovate is the most important cornerstone in solving the environmental problems of the future.

We described an *action plan* for the conversion of a company to a "green" one. As sustainability problems occur everywhere, this change of culture will be demanded by customers and employees alike.

Marketing of environmental aspects is often a challenging subject for a company. The most effective environmental marketing is based on trust. The public is frequently skeptical about environmental claims that lack the support of third-party verification and that may be either incomplete or irrelevant. Companies that follow the necessary environmental marketing principles take a long-term view and do not look only at short-term gains or advantages. The old adage of "say it simple and say it often" is probably reasonable advice, provided that the communication is underpinned with the necessary principles. Tools are available to help companies with this type of environmental marketing.

Many ECODESIGN rules are also available. Choose those rules that best fit the products and culture of your company. An additional benefit is that many if not all of these rules generate cost savings. The important elements are simplified product structure, reduced number of parts, life cycle considerations, and the ability to create innovation leaps.

Production can also not stay as it has been: Energy reduction is required, old product production lines have to be substituted by ones more environmentally benign. The trend is to know completely which substances are used and what is supplied to customers.

Management plays the most important role in the introduction of Ecodesign. Is there a systematic approach? Are ambitious targets set? Is there a review of progress? How is success measured? Several proposals were made in this book and it will be interesting to see what will be really successful worldwide.

Some *examples* were provided for the practical assessment of the Ecodesign directive requirements, PCF, and for the presentation of some practical design principles. Please look at the many tools added – especially in the *checklists*, which should help you to find your own way of managing and implementing ECODESIGN.

What cannot be solved with Sustainability or Ecodesign procedures are political decisions. During times of financial crises, political decision-makers are often not the most fearless, even though these times of crisis are also often the most opportune to effect major change. In the Chinese alphabet, the character for crisis and chance is the same. It is simply a matter of what view one takes.

The chances are there to leave old paradigms behind and to start a new way of doing business, profiting from less energy and less resource consumption, while providing products of excellent environmental performance. The markets are becoming more and more receptive.

The combined activities of gaining independence from fossil-based energy and reducing the environmental impact of products can be also seen as a "job engine" for the future.

In today's world, we can already examine the life cycle of any product and take into account the services and added value related to the product. The logical next step would be the application of the systems in which the products are applied. If such a view is complete, one could generate value by establishing take back systems.

For the future, whole systems where products are applied will need to be considered. At the moment, a system like the "internet" is a sum or a collection of many individual hardware products usually only developed for special purposes such as printing and connected by software for another special task called communication. If these very complex systems and their consequences can be better described, it should be possible to develop much more integrated hardware products *and* through that, saving considerably more resources than looking at the product in isolation.

Besides the lack of Ecodesign rules for such complex structures, new environmental management systems are required for more systematic control of environmental aspects. One example for such an environmental management task will be to limit the tremendous increase of energy consumption of a system like "Internet".

In the future, business will most likely be more about creating value for customers rather than selling pieces of hardware. Nevertheless, products will remain important. How to better manage the process of developing and marketing products with better environmental performance was the task of this book.

We would like to thank you for reading "ECODESIGN – the competitive advantage". We certainly hope that one or more of the various concepts, processes and ideas will inspire you to take action in your own company and to have *Ecodesign become your competitive advantage*.

Ecodesign–The Competitive Advantage

Annexes and Supporting Documents

The European Commission has installed a homepage about life cycle thinking, within which is a broad range of information about the tools available for Ecodesign [1]. Covered are not only tools but also databases for qualified LCA data, a very necessary requirement for all designers. Also, recommendations are provided for different positions in a company, such as environmental manager or designer [2].

In this book, we often describe more simplified and easy to apply methods or even those which are not yet available on the homepage of the EU.

Annex 1 (to Sections 2.3/2.4)

Checklist for the reduction of energy consumption of an E&E product by improved software

1. Are energy saving elements integrated in the software, such as automatic pull-down to stand-by?
2. Are loading commands to batteries, capacitors, etc., checked for energy consumption?
3. What standards are available to estimate comparable modern energy consumption for the equipment like, e.g., Energy Star?
4. Is the energy consumption of the upgraded equipment much higher with the upgraded reused software and how can it be reduced?
5. Is the run time for some tasks too long?
6. Is a combination of new hardware and software necessary to reduce energy consumption, e.g. a combination of switchable power supplies?
7. Are the energy impacts of peripheral components like printers checked for energy consumption?
8. Can hardware in the newly developed product be substituted by software, e.g. fax, now usually integrated?
9. Are there recommendations available for an energy saving (software) mode of the product?
10. Is the possibilities of energy consumption control of a product by software used (e.g. via a control unit together with a motor, pump, heating, etc., to accommodate their optimized working conditions)?

Annex 2a (to Sections 2.3, 2.4 and 4.3)

Checklist of design strategies over the life cycle [3]

Life cycle stage	Strategy	Target
Raw material	**Use alternative materials:** Different materials have a varying impact on the environment in their extraction, recycling, or disposal. The requirement for resources and energy also varies depending on the material selected. Replace those materials of a product that have been identified as particularly resource-consuming (such as virgin aluminium, copper, carbon fiber, etc.) by other materials.	Reduction of environmental impact by using environmentally sound materials, recycled materials, renewable materials, etc.
	Use less of a given type of material: An improvement of the environmental impact can generally be realized by reducing material inputs. Minimize the number of materials that cause additional resource consumption.	Reduction of number of materials by design aiming at optimum strength, integration of functions, etc.
Manufacture	**Use less energy and material:** Different methods of manufacture vary as to their environmental impact. They require different amounts of resources and energy to produce a given product. Environmental damage often results from the application of inadequate production processes. Reduce the amount of energy and material necessary for production. Use alternative types of energy. Avoid or reduce the consumption of auxiliary and process materials in the production process?	Reduction of energy consumption throughout production by means of optimized processes, renewable energy, etc. Reduction of environmental impact caused by consumption of process materials in production process (closed cycles, etc.).
	More efficient use of materials: Optimum use of materials in the production process also helps to reduce costs for the procurement of materials as well as for waste disposal. Reduce waste and/or emissions generated in the production process?	Reduction of waste in production through material efficiency, recycling, etc.
	Avoid joints not allowing easy disassembly: Components for service, maintenance can be gathered. Cost can be also saved for assembly.	Easy disassembly if components should be reused.

Life cycle stage	Strategy	Target
Distribution	**Purchase of external materials/ components:** The environmental quality of a product also depends on the quality of the parts and components purchased from other manufacturers. Procure materials, parts, and components in such a way as to ensure environmentally sound manufacture. Reduce the requirement for transportation in the procurement of external parts.	Procurement of environmentally sound product parts (energy saving, production processes, hazards, etc.).
	Change packaging: As packaging material is useful only for a limited period of time (unless it is returnable) the type and quantity of material used for packaging should be optimized. Especially with products that have to be transported over long distances, the weight of the packaging material, too, has a great influence on the overall consumption of resources. Reduce the weight of the packaging material (often no packaging is required for transportation). Use returnable packaging or renewable or recycled materials for packaging?	Optimization of packaging by taking into account material characteristics, renewability, closed cycles, etc.
Product use	**Realize a high degree of functionality** Reliable and functionally optimized products ensure maximum benefit from the resources used and guarantee customer satisfaction. Improve the functional quality of the product (and its components). Prolong the useful life of the product by means of regular tests of its functionality and operational safety?	Improved functionality by means of upgrading, multi-functionality, etc. (as long as the environmental impact doesn't exceed limits). Improving maintenance through wear detection, remote control, etc.
	Ensure safe use of the product: With products that are used intensively, operational safety is particularly important. Does the product hold a potential risk to the environment. Take measures that minimize this risk.	Avoiding waste during product use. Avoiding risks.
	Reduce energy and material input at use stage: The more often a product is used, the greater the proportion of resources that are used up and waste and emissions generated at the use stage in relation to the overall environmental impact caused by the product during its life cycle.	Reducing the consumption of energy and process materials during product use. Reducing the amount of waste and emissions generated during the use of the product.

Life cycle stage	Strategy	Target
	Use the product as intensively as possible; make intensive use of resources:	Improved usability of products through adaptability, ergonomics, etc.
	Optimum utilization of the product ensures an efficient use of the valuable resources contained in the product.	Improved functionality by means of upgrading, multi-functionality, ...
	Improve handling, functionality, and thus the overall functional quality of the product (and its components). Products that are easy to handle and that require little maintenance, make for efficient use.	Improving maintenance through wear detection, remote control etc.
	Extend the service life by means of regular tests of the functionality and operational safety of the product.	
	Use the product longer; longer use of resources	Durability through dimensioning, surface design, etc.
	A long service life of the product also ensures efficient use of the individual parts and components. Ease of repair prevents premature disposal of the product.	Improving access to, disassembling, and exchange, etc., of parts.
	Prolong the service life of the product also through reuse of components. (A too high energy consumption compared to new products in the market can limit the application.)	
	Environmentally compatible documentation:	Involve consumers.
	Use paper free-instructions or, if necessary, environmentally compatible paper. Inform about environmentally compatible use and take-back or recycling.	
End-of-life	**Disassembly and Recycling**	Organize product take back and ease of disassembling (fastness, etc.).
	The more parts and components of the product can be reused and the easier materials can be recycled, the less material has to be disposed of; the input necessary for disposal will be limited and resources contained in the product may be reused or recycled. Design should aim at reusing parts and components as this approach avoids the destruction of the structure of components and thus preserves the value of individual parts.	Organize reuse of parts (access, remanufacturing, etc.), also reuse of qualified as new components in new products. Organize recycling of materials (separation, labeling etc.).
	Recycle materials (together with recycler).	

Annex 2b (to Sections 2.3, 2.4 and 4.3)

Design checklist for environmentally compatible plants[a] and projects

Phase	Strategy	Target
Planning	Estimate the environmental impact over the whole life cycle of a plant. Derive targets for (reduced) impact for new development or the planned plant.	Reducing environmental impact.
	- Determine the essential inputs and outputs for installation, application, and plant shuttering.	
	- Identify those elements/components of the plant which are the energy consumption (environmental impact) drivers (ovens, fans, heating, cooling, energy intensive processes, etc.) and determine their improvement potential and develop a plan to achieve it (also together with supplier); plan requirements to suppliers.	
	- Check software for unnecessary or energy-consuming commands (loading, no stand-by).	
	- Apply the necessary legal and normative requirements and tendencies.	
	Integrate all relevant stakeholder requirements:	Integration of all necessary requirements.
	- Customer requirements and expectations	
	- Locally important influences (e.g. political demands, investors, NGOs, allowances, etc.).	
	Determine by which environmental properties can advantages in competition be achieved. Show ecological benefit for the customer.	Marketing of advantages.
Procurement and production	For procurement and manufacturing of components and production consumables, select the most compatible solutions possible (frame conditions) and document their application:	Reducing environmental impact.
	- Apply high efficient components produced according to IEC 62 430.	
	- Install efficient energy reuse (waste heat recovery, break energy recovery).	
	- Avoid losses and stand-by.	
	- Observe local/regional restrictions for component and material selection.	
	- Apply environmentally compatible packaging and transport (material, logistics).	

Phase	Strategy	Target
Installation	Apply EH&S concept for plant installation.	Avoidance of environmental impact.
Application	Design plant modular, upgradable, extendable. - Plan possibility of retrofit - Enable upgrades/modifications - Allow reuse of components	Easy service.
	Guarantee energy efficient run of the plant with all possible states of the plant through technical measures and system optimization. Apply, for example, efficient variable speed drives, system and load control, energy management, intelligent power supply, emission management, etc.)	Minimized impact during use.
	Reduce consumption of consumables through technical measures and system optimization.	Reducing consumables.
	Minimize emissions (substances, noise, radiation) during the use of the plant.	Minimizing emissions.
	Inform applicants about the economic advantages of environmentally compatible use (energy, water, consumables, maintenance, etc.) within the instruction manual and, if necessary, offer special training.	Information of applicants.
	Develop a concept for the environmentally compatible recovery of the waste or disposal of unavoidable waste. Include a recycler market analysis, technical possibilities of recyclers, storage of spare parts, and spare parts concept.	Minimizing waste during use.
Take back and recycling	Demonstrate concept for the treatment of shuttered plants (reuse of plant/components, recovery, disposal) with estimation of potential costs and show economic advantages.	Environmentally compatible treatment.

[a]Definitions:

Plant: In an industrial plant, the combination and integration of goods and services under one overall responsibility, a functioning system (industrial plant) to enable a process which covers different, but linked process steps (Translation from a German Wikipedia definition).

Product: Any goods or services [4].

Annex 3 (to Section 2.5)

Costs to be reported by an environmental expert annually to the board (example):
(in close cooperation with other experts like from quality, training, health and safety, purchasing)

1. Preventive costs:
- Training,
- Planning (environmental targets, energy saving equipment and measures, together with purchasing department, audit program)
- Requirements for suppliers; supplier audit and evaluation
- Legal approvals; certificates, proofs to official organizations
- Environmental standardization
- Installation of management systems and regular updates
- Internal reporting system for environmental incidents and accidents for production and products
- Preparing the environmental report of the company; management review

2. Analytical costs:
- Process analysis and monitoring (trends, technological evaluation of manufacturing processes)
- Life Cycle Assessments
- Investigations to improve products (especially cooperation with suppliers to reduce energy consumption of production equipment)
- Answering customer questions; actions to the public; answering questionnaires like from rating agencies

3. Environmental costs:

– *Industrial*
- Investment in plants; changes to match new legal requirements
- Costs for purchase of resources: Water and waste water; waste; emissions; energy; subtracted can be: Earnings by waste sales, own power generation; repair of leakages (air compression). Relation to turnover gives a comparable basis to the report before.
- Costs for waste packages; higher storage costs for hazardous production chemicals
- Travel costs (be careful with such costs as the requirements depend on the nature of the business)

– *Product related*

- Material Input/Output ratio; material quantity for manufacturing per unit;
- Packaging quantity per unit, transport costs, recycling costs per unit;
- Accruals for environmental risks (e.g. producer responsibility for recycling)
- Energy consumption per unit/time
- Costs for the organization of take back (batteries, equipment, packaging) and reporting for government, public, and internally for financial auditors

Annex 4 (to Section 2.5)

Risk management (Checklist) before a new law is put into force, such as the substitution or restriction of a material/substance

1. Check which components are affected by the forthcoming legislation.
2. What volumes of products or components are in storage?
3. Send out questionnaires to customers about their requirements for supply? Are there customers who can continue to use the restricted substances in components?
4. Check at which date it is required to have the substitution ready. At which date do customers have to be supplied with the alternatives? Which date does your company have to be ready from the point of view of their own logistics?
5. Send out questionnaires to suppliers to determine when they will be able to provide substitutes for the restricted component. Which components will no longer be supplied? Which cannot be substituted? When do you need the new components from the supplier?
6. Determine what consequences from the questionnaires for your own product portfolio can be expected?
7. What developments are needed in your own company for an approach to the legislation? When does your company have to be ready?
8. What changes are necessary for your logistics, manufacturing, and planning software?
9. Exchange information about unclear legal texts via your manufacturers' association and make inputs to the government for interpretation?
10. Get experience from your own early development to ask government to grant an exemption if an alternative is technically not yet available.
11. Start an internal information campaign (via board) to involve all relevant departments.
12. Don't forget to integrate affiliations in countries abroad! Inform import/export and purchasing departments and employees.
13. Inform customers and suppliers about your readiness. Go public with a success story, if ready.
14. Develop standard texts for legal contracts to avoid misunderstandings or supply with undesired solutions.
15. Develop internally agreed tests for the correct supply in the case of doubt.
16. Start your planning as soon as you are informed about the first draft and if there is a probability for the law to come.
17. Next time start your activities as soon as information is known that some restrictions are planned or discussions are beginning about potential hazards.
18. After a public discussion (assisted by the manufacturers' association), often the legal restriction will be delayed or implemented through voluntary agreements. This will lessen the impact on the company.

Annex 5 (to Section 2.5)

Cost risks with new legislation (Example for a risk calculation)

A new law (e.g. with restriction of a substance or Ecodesign requirements) can be evaluated for a company if potential actions are evaluated on a cost basis. Very rough estimates can be used because many details of the forthcoming law will not yet be fully known. With this information the company can be made aware that the board of directors has to react and politicians can see where corrective actions are required.

Task	Costs
Environmental analysis for the old product	
Questionnaires to suppliers/Information of suppliers about ability to change and when	
Integration in contracts	
Integration of data into the order software and the production software	
New equipment to plan and to purchase	
Development costs per production line	
Documentation costs/publication	
Potential certificates	
Integration in CE mark (see Low Voltage Directive)	

A risk assessment for the company can go one step further, as shown in the following scheme. It evaluates the current situation in context of the draft of the new regulation and evaluates potential measures to improve the situation. On the right side of the table, the risk can be seen by management and reporting at the right intervals will give the right kind of overview to the board.

Calculation scheme for potential costs caused by risks through potential environmental legislation in comparison to effects after measures taken

Risk from legislation (name risk)	Risk before measure (million USD)	Costs of the measure, e.g., person hours per year (in million USD)	Risk after measure (million USD)	Describe background for calculation	Categorize importance for the company, from 1 (high) to 5 (low)	Total estimated risk. Summary for several rows
e.g. ERP-affected business volume (million USD)	Integrate in calculation probability to come for the products			Which products Data basis Potential change in drafted law Reference to paragraphs in legislation		

Annex 6 (to Section 2.5)

Requirements for material management (Checklist)

The influence that a producer can have over his materials is always lessened when nearly everything (components or parts) is purchased and when the added value is very small. In this kind of circumstance, the manufacturer (or in this case, the assembler) of the end equipment is the last one to get all of the material related information together. In order to manage materials effectively in this kind of situation and more generally, there will need to be:

- Information about cost implications.
- Information about hazardous materials and potential risks, including for products still in the storage that may contain restricted substances.
- Systematic phase out of materials with knowledge about consumed quantities and identification of the areas in the company (product, production, service, recycling, etc.) where the risks are.
- Reporting about hazardous substances in production and products to the public (or rating agencies such as the Dow Jones Sustainability Group Index, if required).

Data base and covered area:

Production: Material flows (quantitatively with all materials required for a process and their content). Waste generated. Materials inputs and outputs. Knowledge about the materials: Hazardous, partly restricted, to be substituted, etc. Costs for health investigations of workers. Trends: For restriction of some materials, for prices for purchasing, for safety of employees, for treatment of waste and emissions, cost development.

Product: Material content. Materials to be declared and to be avoided, restricted materials in some countries. Trends for material prices and for restrictions. Generate information lists for the public if needed for different purposes like information for the public about hazardous substances, information for recyclers, information about noble metals or potentially rare materials internally.

Software: Internally and externally compatible systems for all information required by suppliers and for the company's own applications.

Contracts: Include all requirements into contracts.

Storage in production plant: Potential cost for toxic materials (safety). Better planning for all materials purchased. Reduce storage costs by reducing diversity.

Storage of products and components: Products containing restricted materials can no longer be sold after new legislation is put into force (e.g. RoHS). Mark critical components/ products and sell in time. Extend logistic software to include all marking requirements.

Sales: Provide a list of hazardous substances to inform customers and the public, including statements about what is really not in the product, and what will be avoided in the future.

Service: Components available for repair: New ones, requalified ones, those put on the market before 1.7.2006 (RoHS).

Waste: Costs for take back and recycling; earnings for reuse and recycling of valuable materials.

Define **reporting** for these figures. Investigate risks and define a phase out program for the materials of risk. Check quantities in relation to waste. Reduce resource volumes and waste. Review these figures annually.

Annex 7 (to Section 4.1)

The 12 principles of green engineering [5]

Result of "Defining the principles conference", May 2003, Sandestin

1. Designers need to strive to ensure that all material and energy inputs and outputs are as inherently non-hazardous as possible.
2. It is better to prevent waste than to treat or clean up waste after it is formed.
3. Separation and purification operations should be a component of the design framework.
4. System components should be designed to maximize mass, energy, and temporal efficiency.
5. System components should be output pulled rather than input pushed through the use of energy and materials.
6. Embedded entropy and complexity must be viewed as an investment when making design choices on recycle, reuse, or beneficial disposition.
7. Targeted durability, not immortality, should be a design goal.
8. Design for unnecessary capacity or capability should be considered as design flaw. This includes engineering "one size fits all solutions", etc.
9. Multi-component products should strive for material unification to promote disassembly and value retention (minimize material diversity).
10. Design of processes and systems must include integration of interconnectivity with available energy and material flows.
11. Performance metrics include designing for performance in commercial "after-life".
12. Design should be based on renewable and readily available inputs throughout the life cycle.

Annex 8 (to Section 4.5)

Management system for assessing conformity with the Ecodesign Directive
(referred to in Article 8 of the Ecodesign directive)

1. This Annex describes the procedure whereby the manufacturer who satisfies the obligations of point 2 of this Annex ensures and declares that the energy related product (ERP)[1] satisfies the requirements of the applicable implementing measure. The declaration of conformity may cover one or more products and must be kept by the manufacturer.
2. A management system may be used for the conformity assessment of an ERP provided that the manufacturer implements the environmental elements specified in point 3 of this Annex.
3. Environmental elements of the management system

This point specifies the elements of a management system and the procedures by which the manufacturer can demonstrate that the ERP complies with the requirements of the applicable implementing measure.

3.1. The environmental product performance policy

The manufacturer must be able to demonstrate conformity with the requirements of the applicable implementing measure. The manufacturer must also be able to provide a framework for setting and reviewing environmental product performance objectives and indicators with a view to improving the overall environmental product performance.

All the measures adopted by the manufacturer to improve the overall environmental performance, and to establish the ecological profile of an ERP, if required by the implementing measure, through design and manufacturing, must be documented in a systematic and orderly manner in the form of written procedures and instructions.

These procedures and instructions must contain, in particular, an adequate description of

- The list of documents that must be prepared to demonstrate the ERP's conformity and – if relevant – that have to be made available
- The environmental product performance objectives and indicators and the organizational structure, responsibilities, powers of the management, and allocation of resources with regard to their implementation and maintenance
- The checks and tests to be carried out after manufacture to verify product performance against environmental performance indicators

---

[1]Definition of energy related product (ERP): 'Energy-related product' (a 'product'), means any good that has an impact on energy consumption during use which is placed on the market and/or put into service, and includes parts intended to be incorporated into energy-related products covered by this Directive which are placed on the market and/or put into service as individual parts for end-users and of which the environmental performance can be assessed independently (§2 of Ecodesign directive).

– Procedures for controlling the required documentation and ensuring that it is kept up to date – the method of verifying the implementation and effectiveness of the environmental elements of the management system

3.2. Planning

The manufacturer will establish and maintain

(a) Procedures for establishing the ecological profile of the product;
(b) Environmental product performance objectives and indicators, which consider technological options taking into account technical and economic requirements;
(c) A program for achieving these objectives.

3.3. Implementation and documentation

3.3.1. The documentation concerning the management system should cover the following, in particular:

(a) Responsibilities and authorities will be defined and documented in order to ensure effective environmental product performance and reporting on its operation for review and improvement.
(b) Documents will be established indicating the design control and verification techniques implemented and processes and systematic measures used when designing the product.
(c) The manufacturer will establish and maintain information to describe the core environmental elements of the management system and the procedures for controlling all documents required.

3.3.2. The documentation concerning the ERP will specify, in particular

(a) A general description of the ERP and of its intended use
(b) The results of relevant environmental assessment studies carried out by the manufacturer, and/or references to environmental assessment literature or case studies which are used by the manufacturer in evaluating, documenting, and determining product design solutions
(c) The ecological profile, if required by the implementing measure
(d) Documents describing the results of measurements on the ecodesign requirements carried out including details of the conformity of these measurements as compared with the ecodesign requirements set out in the applicable implementing measure
(e) The manufacturer will establish specifications indicating, in particular, standards which have been applied; where standards referred to in Article 10 are not applied or where they do not cover entirely the requirements of the relevant implementing measure, the means used to ensure compliance
(f) A copy of the information concerning the environmental design aspects of the product provided in accordance with the requirements specified in Annex I, Part 2 (see below).

3.4. Checking and corrective action

(a) The manufacturer must take all measures necessary to ensure that the ERP is manufactured in compliance with its design specification and with the requirements of the implementing measure which applies to it.
(b) The manufacturer will establish and maintain procedures to investigate and respond to non-conformity, and implement changes in the documented procedures resulting from corrective action.
(c) The manufacturer will carry out at least every 3 years, a full internal audit of the management system with regard to its environmental elements.

Annex I of Ecodesign directive, Part 2 Requirements relating to the supply of information

Implementing measures may require information to be supplied by the manufacturer that may influence the way the ERP is handled, used, or recycled by parties other than the manufacturer. This information may include, where applicable

- Information from the designer relating to the manufacturing process
- Information for consumers on the significant environmental characteristics and performance of a product, accompanying the product when it is placed on the market to allow consumers to compare these aspects of the product
- Information for consumers on how to install, use and maintain the product in order to minimize its impact on the environment and to ensure optimal life expectancy, as well as on how to return the product at end-of-life and, where appropriate, information on the period of availability of spare parts and the possibilities of upgrading products
- Information for treatment facilities concerning disassembly, recycling, or disposal at end-of-life

Information should be given on the product itself wherever possible.

This information will take into account obligations under other Community legislation, such as Directive 2002/96/EC.

Annex 9 (to Section 4.5)

Inclusion of Ecodesign elements from IEC 62430 into existing management systems

In ISO 9001, Ecodesign elements can be integrated into the scheme by adding environmental elements to corresponding, and already existing quality elements (in brackets) such as the extension of customer expectations (5.2), or adding environmental targets and product profiles to quality targets (5.4.1). On the other hand, ISO 14001 is more difficult because it doesn´t have a design orientation, and the focus is primarily on the production phase. Therefore, a selection from IEC 62430 should be inserted to address Ecodesign properly. A proposal for integration is made in the ISO 14001 scheme as shown in the following overview.

As it can be seen in ISO 9001, the elements of Ecodesign (e.g. from IEC 62430) fit easily into the elements of product design. In ISO 14001, integration is more difficult as this system is directed to production systems. With the red marked additions, some integration becomes possible. For ISO 14001, all elements can more easily be integrated. With new ISO 14006 guidelines for the integration will be developed.

ISO 9001	ISO 14001

5   Responsibility of management
5.2 Customer expectations
5.4 Planning
    5.4.1 Quality targets
    5.4.2 QM system planning
7   Product realization
7.2 Customer oriented processes
    7.2.1 Definition: customer requirements
    7.2.3 Communication to customers
7.3 Design and development
    7.3.2 Design requirements
    7.3.3-7.3.6 Results, check, verification.
7.4 Purchasing
    7.4.1 Purchasing
    7.4.2 Purchasing information
7.5 Production and service
    7.5.5 Handling ...
	2. Policy, targets, programs
	2.2 Env. targets and programs add
	product related environmental protection
	2.2.2 Special regulations with program e.g.
	    new development project.
	3. Organization and Personal
	3.1.1. Responsibility
	4. Influence on Environment
	4.1 Legislation
	4.2 Evaluation of impact on the
	environment....by products
	4.5 Organization and controlling
	Insert all product related elements here
	4.5.2
	7. Validation

Annex 10 (to Section 4.5.2)

Design steps for Ecodesign according to IEC 62430 with some additional extensions for a system approach

Activity list (acc. to IEC 62430)	Questions (acc. to IEC 62430)	Tools (acc. to IEC 62430)	Description of the product in its application system	Tools for additional investigation
Product´s environmental parameters	What are the elements and life cycle stages of the product?		With life cycle thinking or complete LCA the important elements or stages can be identified. Real impact by the product in the market can only be estimated in total as a sum of all products in the application system (all cellular phones, all washing machines, etc.)	ISO 14040ff. Scenario analysis, includes volumes in the market
Regulatory and market requirements. Needs of customers and stakeholders; relation to environmental aspects to be achieved throughout the life cycle of the product	Who are the stakeholders and what do they expect from the environmental attributes of the products?	Environmentally conscious design (ECD) checklist		
Acquire information from supply chain	What information on relevant life cycle stages is needed (e.g. materials content and energy consumption of components)?		Include info from own production. Include also LCA data directly from supplier	Contracts for exclusion of negative aspects, substances; Software system with material content used for production and in components

Benchmark against the competitor's products	What are the competitor's products environmental strength and weaknesses?	ECD Bench-marking	Benchmarking is also necessary in comparison to environmental targets of governments and necessary environmental requirements in comparison with targets from "Kyoto" etc.	Use official targets, compare with development of sales figures or development of systems (e.g. exponential growth of internet and reduction necessity from that fact)
Identify significant environmental aspects and relevant parameters:- Develop a life cycle flow for the product by selecting appropriate life cycle stages including concepts for end-of-life treatment - Analyze and evaluate the impact on the environment taking into account the foreseeable product life cycle	What opportunities are there to improve environmental attributes of the product? How to coordinate customer needs, benchmarking results and environmental assessment results into common improvement tasks?	ECD Benchmarking Environmental QFD LCT assessment tools	Which fraction of the waste products really can be recovered in the corresponding market? Reuse analysis (If reuse is decided for a more generation product planning is necessary) Compare environmental aspects with all other aspects of the product including expected costs to find trade-offs Do trade-off analysis also between the detected environmental aspects	Market (recovery) analysis with available recyclers; also check technology shredder vs. disassembly Cost benefit/analysis of components/materials (cf. also IEC 62309 "qualified as good as new") Life cycle costing; Design to cost EQFD EFMEA Trade-off check

(continued)

Design steps for Ecodesign according to IEC 62430 with some additional extensions for a system approach (continued)

Activity list (acc.to IEC 62430)	Questions (acc. to IEC 62430)	Tools (acc. to IEC 62430)	Description of the product in its application system	Tools for additional investigation
- Compile the result of the environmental analysis and stakeholders' requirements				
Define environmental targets	What are the technical, economical and business issues and feasibility associated with proposed improvements? What are the specific tasks and resources for achieving the environmental targets?			

After defining product-related environmental targets, the components will need to be evaluated to determine how changes to each component can contribute to the overall target. Allocations to the product structure should be made according to the procedures described in Sections 3.3 and 4.5. As an example: If a structure is modular, with each module containing further components, and the energy reduction target is 50%, one needs to determine in which modules the highest reduction potential is possible. If a new motor controller can contribute 45% of the reduction, the rest and perhaps more can be most likely found by examining reduction possibilities in other modules, such as heating elements or power supplies. If the reduction potential at the module level cannot directly be identified, an analysis at the component level will help find those components where improvement should start.

Annex 11: List of Mentioned Standards

BSI

BSI PAS 2050:2008; Specification for the measurement of the embodied greenhouse gas emissions in products and services

IEC

IEC Guide 113:2000; Materials declaration questionnaires – Basic guidelines

IEC 60300-3-3:2004; Dependability management – Part 3-3: Application guide – Life cycle costing

IEC PAS 61906:2005; Procedure for the declaration of materials in products of the electrotechnical and electronics industry (future ed. in IEC 62 474)

IEC 62309:2004; Dependability of products containing reused parts – Requirements for functionality and test

IEC 62430:2009; Environmentally conscious design for electrical products and electronic products systems

ISO

ISO 9001:2008; Quality management systems – Requirements

ISO 14001:2004; Environmental management systems – Requirements with guidance for use

ISO 14020:2000; Environmental labels and declarations – General principles

ISO 14021:1999; Environmental labels and declarations – Self-declared environmental claims (Type II environmental labeling)

ISO 14024: 1999; Environmental labels and declarations – Type I environmental labeling – Principles and procedures

ISO 14025:2006; Environmental labels and declarations – Type III environmental declarations – Principles and procedures

ISO 14040:2006; Environmental management – Life cycle assessment – Principles and framework

ISO 14041:1998; Environmental management – Life cycle assessment – Goal and scope definition and inventory analysis (see also: ISO 14 042-44)

ISO 14044:2006; Environmental management – Life cycle assessment – Requirements and guidelines

ISO 14064:2009; Green House Gases Part 1; Specification with guidance at the organization level for quantification and reporting of greenhouse gas emissions and removals, Part 2; Specification with guidance at the project level for quantification, monitoring and reporting of greenhouse gas emission reductions or removal enhancements, Part 3; Specification with guidance for the validation and verification of greenhouse gas assertions

ISO WD 14067:2009; Carbon footprint of products; Part 1; Quantification, Part 2; Communication

ISO TR 14062:2002; Environmental management – Integrating environmental aspects into Product Design and Development

ISO 22628:2000; Road vehicles – Recyclability and recoverability – calculation method

OSHA

OHSAS 18001:1999; Occupational Health and Safety Assessment Systems: Specifications

References

Frontmatter

1. The Boston Consulting Group GmbH & Partner: Vision und Strategie (VI), Spielregeln ändern, Kommentare, München, Düsseldorf, Zürich, May 1991.
2. http://eur-lex.europa.eu/LexUriServ/site/en/com/2001/com2001_0068en01.pdf

Chapter 1

1. International Energy Agency, World Energy Outlook 2008, OECD/IEA, 2008
2. Christian Hagelüken, 12.04.2007, Umicore Precious Metals Refining
3. U.S. Geological Survey, Mineral Commodity Summaries, 2007
4. The Economics of Climate Change: The Stern Review, Cambridge University Press 2007
5. http://www.eiatrack.org
6. Directive 2002/95/EC of the European Parliament and of the Council of 27 January 2003 on the restriction of the use of certain hazardous substances in electrical and electronic equipment; OJ 13.2.2003; L37/19–23
7. Ecodesign directive: Directive 2009/125/EC of the European Parliament and of the Council of 21 October 2009 establishing a framework for the setting of ecodesign requirements for energy related products (recast) OJ 31.10.2009, L285/10 EuP directive: Directive 2005/32/EC of the European Parliament and of the Council of 6 July 2005 establishing a framework for the setting of Ecodesign requirements for energy-using products and amending Council Directive 92/42/EEC and Directives 96/57/EC and 2000/55/EC of the European Parliament and of the Council, OJ 22.7.2005; L191/29
8. http://ec.europa.eu/enterprise/newapproach/legislation/guide/document/introduction.pdf
9. http://ec.europa.eu/environment/eussd/escp_en.htm
10. http://ec.europa.eu/enterprise/newapproach/legislation/guide/document/chap07.pdf
11. http://eur-lex.europa.eu/LexUriServ/site/en/com/2001/com2001_0068en01.pdf
12. http://ec.europa.eu/environment/ipp/mobile.htm
13. UNFCCC, http://unfccc.int
14. http://cdm.unfccc.int/index.html
15. http://unfccc.int/2860.php
16. http://www.emissions-trading-info.com
17. http://unfccc.int/files/meetings/cop15/application/pdf/cop15_cph_auv/pdf
18. http://www.epa.gov/captrade
19. http://ec.europa.eu/environment/climat/climate_action.htm
20. ISO 14001:2004; Environmental management systems – Requirements with guidance for use

21. http://ec.europa.eu/environment/index_eu.htm
22. ISO 14020:2000; Environmental labels and declarations – General principles
23. ISO 14021:1999; Environmental labels and declarations – Self-declared environmental claims (Type II environmental labelling)
24. ISO 14024:1999; Environmental labels and declarations – Type I environmental labelling – Principles and procedures
25. ISO 14025:2006; Environmental labels and declarations – Type III environmental declarations – Principles and procedures
26. http://www.eccj.or.jp/top_runner/index.html
27. ISO 14040:2006; Environmental management – Life cycle assessment – Principles and framework
28. ISO WD 14067:2009; Carbon footprint of products; Part 1 Quantification, Part 2 Communication
29. Directive 2005/32/EC of the European Parliament and of the Council of 5 April 2006 on energy end-use efficiency and energy services and repealing Council Directive 93/76/EEC; OJ, 27 April 2005: L114/64
30. http://hse-rohs.oeko.info/fileadmin/user_upload/Documents/RoHS_Hazardous_Substances_Final_Report.pdf
31. http://eur-lex.europa.eu/JOHtml.do?uri=OJ:L:2007:136:SOM:EN:HTML; http://echa.europa.eu/home_en.asp
32. Directive 2002/96/EC of the European Parliament and of the Council of 27 January 2003 on waste electrical and electronic equipment (WEEE) – Joint declaration of the European Parliament, the Council and the Commission relating to Article 9; OJ 13.2.2003: L37/24
33. Directive 2008/98/EC of the European Parliament and of the Council of 19 November 2008 on waste and repealing certain Directives, OJ 22.11.2008: L312/3
34. www.nagpi.net
35. Buy It Green Network – www.iclei-europe.org/big-net
36. Green Purchasing Network – www.gpn.jp/English/
37. Claude Fussler and Peter James, Driving EcoInnovation, Pitman Publishing, 1996 ISBN 0 273 62207 2
38. IEC 60303-3-3:2004; Dependability management – Part 3-3: Application guide – Life cycle costing
39. ZVEI White paper on energy efficiency

Chapter 2

1. Chan Kim W and Renee Mauborgne. "Knowing a Winning Business Idea When You See One". *In* Harvard Business Review, September–October 2000
2. Wimmer W. et al. 'Ecodesign implementation: A systematic guidance on integrating environmental considerations into product development', Springer 2004
3. IEC 62430:2009; Environmentally Conscious Design for Electrical and Electronic Products
4. Mathis Wackernagel and William E. Rees, 1996, "Our ecological footprint: reducing human impact on the earth, New society publishers
5. Song JS, 2009, Development of low carbon product design system using embedded greenhouse gas emissions, Ph D Thesis, Department of Environmental Engineering, Ajou University
6. IPCC, 2006, IPCC Guideline for national GHG inventories
7. WRI/WBCSD, 2004, GHG Protocol; A Corporate Accounting and Reporting Standard
8. ISO 14064:2006; Green House Gases; Part 1: Specification with guidance at the organization level for quantification and reporting of greenhouse gas emissions and removals; Part 2: Specification with guidance at the project level for quantification, monitoring and reporting of greenhouse gas emission reductions or removal enhancements; Part 3: Specification with guidance for the validation and verification of greenhouse gas assertions

9. ISO/WD 14067:2009; Carbon footprint of products; Part 1 Quantification, Part 2 Communication
10. BSI PAS 2050:2008; Specification for the measurement of the embodied greenhouse gas emissions in products and services
11. ISO/WD 14067:2009; Carbon footprint of products; Part 1 Quantification
12. IPCC, 2007, 'Climate change 2007: The physical science basis'
13. EC, 2008, commission regulation No 1275/2008, Implementing directive 2005/32/EC with regard to Ecodesign requirements for standby and off mode electric power consumption of electrical and electronic household and office equipment
14. Lee, KM, 2008, The GHG emission factors from the conversion of the Korean LCI database
15. Regulation (EC) No 1907/2006 of the European Parliament and of the Council of 18 December 2006 concerning the Registration, Evaluation, Authorisation and Restriction of Chemicals (REACH), OJ 31.12.2006; L 396, 1–849
16. http:// digitaleurope.org
17. http://www.ZVEI.org
18. ZVEI: http://zvei.org/index.php?id=2271; Guide for Umbrella Specs: https://www.zvei.org/fileadmin/user_upload/Fachverbaende/Electronic_Components/Umbrella_Specs/UmbrellaSpecs_Version2_1.pdf
19. IMDS system: http://www.mdsystem.com/html/en/home_en.htm
20. List of available components in electroindustry see: ZVEI White paper 2008: Generating distributing and using energy intelligently, http://en-q.de/whitebook.html
21. ISO 22628:2000; Road vehicles – Recyclability and recoverability – calculation method
22. Boothroyd, G, Dewhurst, P,1994, Product design for manufacture and disassembly; Marcel Dekker, Amsterdam http://www.dfma.com
23. IEC 62309:2004; Dependability of products containing reused parts – Requirements for functionality and test
24. For Medical industry see: COCIR 2007, Greenpaper "Good refurbishment practice", http://www.cocir.org/uploads/documents/COCIR-GRP-Green-Paper-version-1–22%20Nov-2007.pdf
25. Graedel TE, Allenby BR, 1995, "Industrial Ecology", Prentice Hall, Englewood Cliffs, New Jersey pp 126
26. IEC 62430:2009; Environmentally Conscious Design for Electrical and Electronic Products
27. ISO 14044:2006; Environmental management – Life cycle assessment – Requirements and guidelines
28. EU Commission, 2005 directive 2005/32/EC 'Establishing a framework for the setting of Ecodesign requirements for Energy using products (Eup)

Chapter 3

1. http://w1.siemens.com/annual/08/en/portfolio/environmental_portfolio.htm
2. Keizo, Fujimori: Publications of the European Japan Experts Association, Vol. 2 (1997) pp 127–131
3. http://www.sony.net/SonyInfo/Environment/activities/theme/index.html
4. http://www.sustainability-indexes.com/07_htmle/sustainability/corpsustainability.html
5. The Concise Oxford Dictionary of Current English, Sixth Edition, Oxford: Clarendon Press, 1976
6. Roper Starch Worldwide (1997), Green Gauge Report, Roper Starch Worldwide Inc., New York, NY
7. www.terrachoice.com
8. Lee, Kun-Mo and Uehara Haruo, APEC, 2003; "Best practices of ISO 14021: Self-declared environmental claims"
9. Improvement strategies taken from the tool ECODESIGN PILOT – www.ecodesign.at/pilot
10. Environmental Product Declaration of the digital Pocket Memo – www.ecodesign-company.com/solutions

11. WRI/WBSCD, calculation tool for direct emissions from stationary combustion, version 3.0, 2005
12. World resources institute, The greenhouse gas protocol: a corporate accounting and reporting standard, 2004
13. BSI PAS 2050:2008; Specification for the measurement of the embodied greenhouse gas emissions in products and services
14. IEC 60303-3-3:2004; Dependability management – Part 3-3: Application guide – lifecycle costing
15. http://en.wikipedia.org/wiki/Activity-based costing
16. http://www.epa.gov/dfe/pubs/garment/lcds/micell.htm
17. http://en.wikipedia.org/wiki/Lotus_effect or ultra sonic washing:
18. IEC 62309:2004; Dependability of products containing reused parts – Requirements for functionality and tests
19. ISO 9001:2008; Quality management systems – Requirements
20. Scheer, August-Wilhelm: Aris. Business process frameworks, 3rd edition, Springer, Berlin 1999
21. Low Voltage Directive: 2006/95/EC of the European Parliament and of the Council of 12 December 2006 on the harmonisation of the laws of Member States relating to electrical equipment designed for use within certain voltage limits (codified version), OJ 16.01.2007; L374/10–19
22. GRI, http://www.globalreporting.org/Home
23. Wimmer, W. et al. ECODESIGN Implementation: A systematic guidance on integrating environmental considerations into product development, Springer 2004

Chapter 4

1. "Green Engineering Principles": declared on the Sandestin conference "Defining the principles" in Sandestin, Florida, 18–23 May 2003
2. BSH: The Protos stove for developing countries, http://www.bsh-group.com/index. php?page=109906
3. http://baumev.de
4. Arnt Meyer, 'What's In It For The Customers? Successfully Marketing Green Clothes', Business Strategy and the Environment (Vol. 10, 2001), pp. 317–330 at 320
5. Patrick Hartmann, Vanessa Apoalaza Ibáñez and F. Javier Forcada Sainz, 'Green Branding Effects on Attitude: Functional Versus Emotional Positioning Strategies', Marketing Intelligence & Planning (Vol. 23, No. 1, 2005), pp. 9–29 at 11
6. GEN – website is www.globalecolabelling.net
7. Wimmer et al. ECODESIGN Implementation, A systematic guidance on integrating environmental considerations into product development, Springer 2004
8. Kun Mo LEE, The RoHS Manual for SMEs, Asian Productivity Organization, 2008
9. IEC PAS 61906:2005; Procedure for the declaration of materials in products of the electrotechnical and electronics industry
10. IEC 62430:2009; Environmentally Conscious Design for Electrical and Electronic Products; appendix Figure B
11. Quella, F.: Umweltverträgliche Produktgestaltung, Verlag Publicis MCD, Erlangen-München 1998; Quella, F.: Environmentally Compatible Product Design – How it works with the Siemens Standard SN 36350, In "Defining the principles", Destin, Florida 18–23 May 2003
12. EFQM; http://www.efqm.org
13. Melzer, K.: Integrierte Produktpolitik bei elektrischen und elektronischen Geräten zur Optimierung des Product-Life-Cyle (Hrsg. K.Feldmann, M. Geiger, Fertigungstechnik Erlangen Band 169, Meisenbach Verlag Bamberg, 2005

14. WOIS – Contradiction oriented innovation strategy; http://www.wois-innovation.de
15. Bionik: Network of competence, http://www.biokon.net·
16. www.ecodesign.at/pilot
17. Steelcase Inc.

Chapter 5

1. Directive 2006/42/EC of the European Parliament and of the Council of 17 May 2006 on machinery, and amending Directive 95/16/EC (recast); OJ 9.06.2006; L157/24
2. EC 2000, Guide to the implementation of directives based on the New Approach and the Global Approach (http://europa.eu.int/comm/enterprise/newapproach/newapproach.htm)
3. Commission Regulation (EC) No 107/2009 of 4 February 2009 of the European Parliament and of the Council with regard to ecodesign requirements for simple set-top boxes; OJ 5.2.2009: L36/13
4. www.ecocomplexstb.org/request_form_18.php
5. Council decision of 22 July 1993 concerning the modules for the various phases of the conformity assessment procedures and the rules for the affixing and use of the CE conformity marking, which are intended to be used in the technical harmonization directives OJ 30.08.1993; L 220, 0023–0039
6. Mudgal, S. Lyama, S. Turunen, L. Tinetti, B.: Preparatory Studies for Eco-design Requirements of EuPs (II) Lot 18 Complex set-top boxes Final Report, Bio Intelligence Service, European Commission DG TREN, 2007
7. http://ec.europa.eu/enterprise/ecodesign/final_report3.pdf
8. ISO WD 14067:2009; Carbon footprint of products-Part 1 quantification
9. IPCC, 2007, Fourth Assessment report: Climate change 2007: The Physical Science Basis, Cambridge University Press
10. Swiss Centre for Life Cycle Inventories, 2009; Ecoinvent Database v 2.1, 1998–2009
11. European Commission DG TREN, 2008; Preparatory studies for Eco-design requirements of EuPs lot 18 Complex set-top-boxes final report
12. ISO/IEC Guide 2, 1996

Backmatter

1. http://lca.jrc.ec.europa.eu/lcainfohub/serviceList.vm
2. http://lca.jrc.ec.europa.eu/lcainfohub/userGuidance.vm
3. www.ecodesign.at/pilot/
4. DIN EN ISO 14021, 2001
5. Sandestin Florida, 2003

Index